0-07-004563-1	Baker	*Downsizing: How to Get Big Gains from Smaller Computer Systems*
0-07-046321-2	Nemzow	*The Token-Ring Management Guide*
0-07-032385-2	Jain/Agrawala	*Open Systems Interconnection: Its Architecture and Protocols, rev. ed.*
0-07-707778-4	Perley	*Migrating to Open Systems: Taming the Tiger*
0-07-033754-3	Hebrawi	*OSI Upper Layer Standards and Practices*
0-07-049309-X	Pelton	*Voice Processing*
0-07-057442-1	Simonds	*McGraw-Hill LAN Communications Handbook*
0-07-060362-6	Spohn/McDysan	*ATM: Theory and Applications*
0-07-042591-4	Minoli/Vitella	*ATM & Cell Relay Service for Corporate Environments*
0-07-067375-6	Vaughn	*Client / Server System Design and Implementation*
0-07-020359-8	Feit	*SNMP: A Guide to Network Management*
0-07-042588-4	Minoli	*Imaging in Corporate Environments*
0-07-005089-9	Baker	*Networking the Enterprise: How to Build Client / Server Systems That Work*
0-07-004194-6	Bates	*Disaster Recovery for LANs: A Planning and Action Guide*
0-07-046461-8	Naugle	*Network Protocol Handbook*
0-07-046322-0	Nemzow	*FDDI Networking: Planning, Installation, and Management*
0-07-042586-8	Minoli	*1st, 2nd, & Next Generation LANs*

Wireless Networked Communications

Wireless Networked Communications

Concepts, Technology, and Implementation

Regis J. Bates

McGraw-Hill, Inc.

New York San Francisco Washington, D.C. Auckland Bogotá
Caracas Lisbon London Madrid Mexico City Milan
Montreal New Delhi San Juan Singapore
Sydney Tokyo Toronto

Library of Congress Cataloging-in-Publication Data

Bates, Regis J.
Wireless networked communications : concepts, technology, and implementation / Regis J. Bates.
p. cm.
Includes index.
ISBN 0-07-004674-3
1. Wireless communication systems. 2. Local area networks (Computer networks) I. Title.
TK5103.7.B39 1994
621.382—dc20 94-10074
CIP

2 3 4 5 6 7 8 9 0 DOC/DOC 9 0 9 8 7 6 5

ISBN 0-07-004674-3

The sponsoring editor for this book was Jerry Papke, the editing supervisor was Kimberly A. Goff, and the production supervisor was Pamela A. Pelton. This book was set in Century Schoolbook. It was composed by McGraw-Hill's Professional Book Group composition unit.

Printed and bound by R. R. Donnelley & Sons Company.

Contents

Preface

When initially considering this book, I was faced with many decisions that kept me guessing: What is it that we really need to know about the various wireless systems that hasn't already been widely published? Who would even care about the use of wireless systems and components? Many other such questions kept going through my mind. Fortunately for me, I wound up in a position where I encountered a group of telecommunications managers and engineers who are normally expected to be on top of the evolution of these components. However, many of these managers and engineers were caught in a bind by the pressure their management exerted on them in an effort to get answers regarding the use of wireless LANs, bridges, and telephone systems. These managers and engineers expressed concern that the only place they could get the information was through the vendor community. This should not have been a problem, except that vendor responses had a distinct sales overture to them. The telecommunications managers and engineers therefore were unsure whether they could rely on the validity of the information or if the representation of the strengths and weaknesses of the systems was accurate. Further, since each vendor had taken a slightly different approach in their implementation of the systems, these users were equally concerned about which approach would "win out" over the others. Making an incorrect selection could prove expensive and embarrassing for these users and their organizations. Thus, I set out to do some additional research on the available products and services out in the real world.

To the industry's chagrin, as I soon discovered, the industry was manufacturing products and services that they thought the user community wanted, without really asking it. *De facto* standards were being implemented by the industry and the regulatory bodies as they wrestled with the best approach toward acceptance and standardization of spectrum to be allocated, products, and services. Recently the

Federal Communications Commission decided to auction off radio spectrum to some 2500 potential licensees for the implementation of personal communications for the future. This should be completed by the time this book is published. The FCC will have raised a potential $10 to $12 billion by this auction. However, there will still be the issue of the many different approaches to getting the most out of the limited frequencies. Should developers use an analog or digital technique? If digital, should the modulation and access be through time division or code division and frequency hopping techniques? These tougher questions will take much longer to answer and as the vendors continue to roll out products and services, these decisions will involve investments that must yield a return. For our suppliers to be in business for the long term, they must be able to expect a return on their investment and make a profit to sustain operations. Given this reality, I found another problem. The products and services that were being offered were extremely volatile in terms of longevity. Many of the products that I researched are no longer offered; others have changed so significantly that users would not consider them the same. This moving target meant I needed to recheck everything twice (or more), yet the changes still kept coming.

Then how can one be sure that a decision to implement or to plan for a wireless complement to their networks will be viable several years later? Clearly, the answer will never come from a single reference, such as this book. This is a dynamic industry. Things change so dramatically that users and vendors must constantly stay on top of the changes that occur, as they occur. However, to get a solid base of the technology, concepts, and possible applications of the wireless communications systems, this book is the best place to start.

I chose to vary the delivery of the information from some technical and some nontechnical approaches. For the novice who picks this book up to learn something about wireless systems and concepts, I have attempted to use analogies and examples that will meet that need. For the semi-experienced and the more experienced user or vendor, I have tried to represent the technical aspects of the design considerations. This makes this book different from the normal publications since they are geared solely toward the technical discussion and the engineering theory behind the systems. These other publications are clearly well done in their representation of the information; many exceptional publications exist and should be the next step in the learning curve for the reader. I personally found that this change of delivery was something for which many users and vendors were looking. Several suggestions from real users, my reviewers, have also been incorporated.

The industry will probably be in a state of flux for the next few years as new products continue to be introduced and the technologies and standards become solidified. During this period of change, the concepts listed in this book will hopefully assist the reader in deciding what to look and to watch out for. The use of wireless communications or components will be an integral part of our networks for many years to come. We should not look to wireless communications as replacements to our wired communications networks but as complementing these networks. Where a wired system will be inefficient, difficult to install, or just plain unsightly, then the wireless option may suffice as a suitable substitute.

Two parallel paths are being taken in the industry: the first is the deployment of fiber optics as close to the business and residential user as possible within financial constraints; the second is the use of wireless communications with some limitations in the capacities that are available, but with fewer limitations from a financial perspective. By the turn of the century, we are likely to see these parallel paths converging into a single entity. The user will not have to be concerned about wired versus wireless connectivity, only about the application they need to address. The either/or connection will be a commodity again, just as the wired telephone services became in the late 1980s. We are in a position to see the networks merge, converge, and emerge as merely bandwidth that satisfies our needs for a price. We should embrace the opportunity to help shape this industry and to experiment with the variations of wireless components, before diving head first into a system or a concept. This will satisfy our interests and help to meet our needs to be accessible at any time, and anywhere.

Regis J. Bates

Acknowledgments

Without the continued support of many friends and vendors (who are also friends), I would still be staring at a blank piece of paper. The ability to garner my thoughts during busy and hectic periods requires that these supporters keep at me and from time to time "kick start" my brain. I am forever grateful for their continued devotion and assistance. Specifically my family and staff and the editorial staff at McGraw-Hill have been extremely patient and accommodating—especially as deadlines come due and I have to hibernate to complete a project.

As I continue to work with industry developers and marketeers, I owe special thanks to them for their openness and their contributions. When I err, they are there to provide support and correctness. The folks at Motorola, SpectraLink, InfraLan, Laser Communications, Ericsson, and Photonics have all been especially helpful in assisting me with the completion of this book. I do appreciate their help and guidance.

Lastly, as I continually bring new deadlines into the fold that stress an already busy staff, they have to be admired and recognized for their efforts. Graphic support from Gabriele Peschkes and typing support from my fellow workers have led to meeting the goal of completing this book.

Without all of you, I would not be producing this acknowledgment. Further, the writing of a book would be the last thing I would be able to accomplish by myself. *Thanks to you all!*

Wireless Networked Communications

Chapter

1

Fundamentals of Wireless Communications

Probably one of the most interesting technologies in industry today is the wireless world. This statement is not founded on the "latest and greatest" in communications breakthroughs. All too often, we hear how evolving technology will unshackle us from the traditional "pairs of wires" which provide our present-day communications. Wireless, however, has been around for decades, in variations of uses and techniques. Only recently have newer applications breathed life back into mundane services, sparking international interest in the applications, bandwidths, and legalities of wireless communications. Today everyone is buzzing about the future of the wireless world and our abilities to communicate in general.

Currently, wireless communications evolve around both the old approaches and newer techniques (see Fig. 1.1). These include:

- Cellular
- Infared
- Meteor burst
- Microwave
- Packetized data over radio
- PCS/PCN
- Satellite
- Specialized mobile radio
- Spread-spectrum radio
- Troposcatter
- Two-way radio

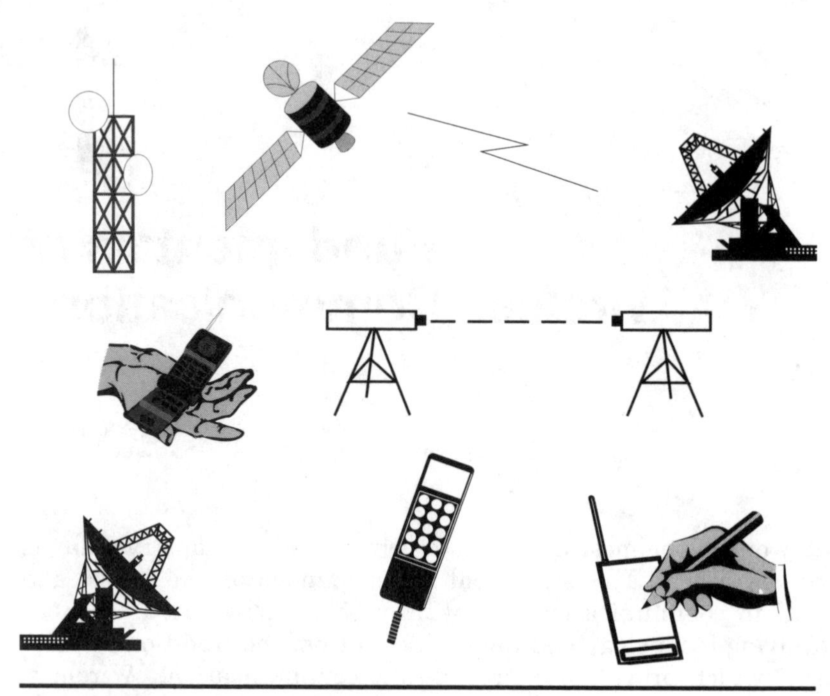

Figure 1.1 Some of the available wireless tools.

We'll examine each of these in more detail to gain an understanding of the technologies and capabilities involved.

History of Wireless Communications

Wireless communications are nothing new. Since the early days of civilization, various forms of communication took place without the advantage of physical connectivity. In tribal jungle environments, drums were a primary means of communicating. As message senders beat on either drums or hollowed-out logs, the reverberating sounds were interpreted at the other end. In many situations, the drumbeat would travel only a limited distance, so various relay points were needed. The first receiver would acknowledge the sender's message through a series of return drumbeats, then relay the same message to the next receiver. This limited form of transmission met the need but was subject to a lot of noisy interference and misunderstanding. Therefore, the message was sent over and over again to minimize potential errors in interpretation. Crude, but it worked.

In early American times, native American tribes used smoke signals as a limited-distance form of communication. Drawbacks to smoke signals included distance limitations based on line of sight, a limited alphabet, and errors caused by the wind. If the smoke puff was blown away or dissipated too soon, the communication was lost.

The introduction of the semaphore flag deepened the scope of communication through an enhanced alphabet that could address the language needs, but was also limited to line-of-sight daytime operation. Hardly a reliable or widely available capability, it was effective in certain circumstances. Since semaphore signaling allowed a full alphabet to be used, messages could be more extensive and detailed. Drawbacks were the limited distance (requiring constant relay of the information), the added time needed to send detailed messages, and the risk of undetected interception of the message.

In the nineteenth century, light beams were used for short-haul communications, particularly in military contexts. Very detailed messages could be transmitted by a coded sequence (Morse code) of blinking lights from sender to receiver. Again, this was effective over limited distances and provided a quiet, yet visible means of communication. Drawbacks included limited distance, unauthorized reception of information due to visibility at various angles, and risk of interception. Security was always suspect, so a form of alphabetic encryption was introduced as a safeguard. This required an ever-changing code set, along with special handling and extra time to manually decipher the transmitted message. Further, the cipher code had to be kept current at all locations so that correctness could be achieved.

Radio-Based Systems

As radio-based systems emerged, wireless communications became more readily available and easier to use. An electric transmitter was used to reproduce sound waves and modulate human speech onto a baseband radio frequency. The radio wave carrying the transmitted signal could travel greater distances, allowing far more reliability and minimizing the relay process. Several portions of the radio frequency spectrum were allocated (assigned) to these transmission systems. Each carried its own particular capacities and had distance limitations associated with the band (range of frequencies) used for transmission.

In radio transmission, human speech must first be converted to an electrical signal. This signal is analogous to the composition of the sound-pressure changes produced by the human voice. Hence, the term *analog communications*. The analysis of sound waves is a key part of radio communications theory. Knowledge of radio principles is critical to understanding how the various wireless communications techniques function. In early telecommunications systems (particularly the telephony world), radio was an integral part of network development. As newer systems emerged, modifications and enhancements enabled networks to carry all forms of communication, including voice, data, telegraph, image, fax, and video.

Free-space communication

Radio systems propagate information in free space. This free-space communication obviates some of the problems faced by other transmission systems. For example, wired systems require a physical medium and are difficult to install in certain geographic areas. Advantages of the radio system include the ability to:

- Span bodies of water, such as lakes or rivers, where a cable facility would require special treatment to prevent seepage onto the copper conductors
- Overcome transmission obstacles posed by mountains and deep valleys, where cable costs would be prohibitive to install and difficult to maintain
- Bypass the basic interconnection to the local telephone provider (telco) or post telephone and telegraph (PTT) company

Figures 1.2 to 1.4 summarize these obstacles and show the advantages afforded by radio-based systems.

In Fig. 1.2, the local service provider realized that using a cable-based system would require an underwater run of wiring which would be prone to water leakage. An alternative was an overwater (aerial) pull requiring extensive cable support systems (including pole construction with guyed wires) that would be prone to wind destruction and cable deterioration. Figure 1.3 depicts the use of radio systems to overcome the terrain problems associated with a mountain or valley. Rather than construct a cable system rising over the mountain, the local provider used a radio system as a relay point.

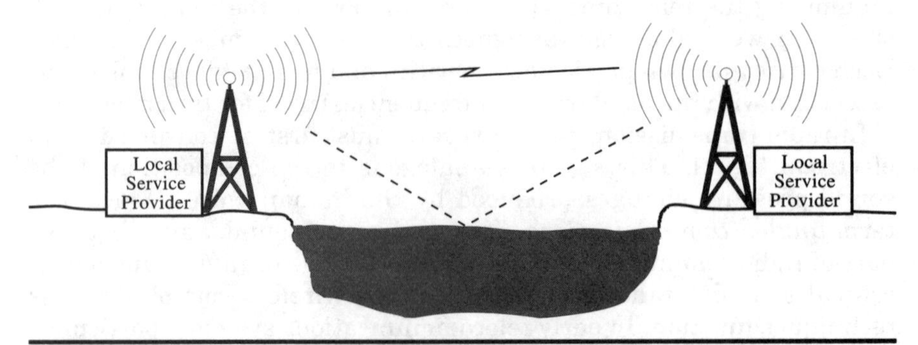

Figure 1.2 Radio used to span bodies of water.

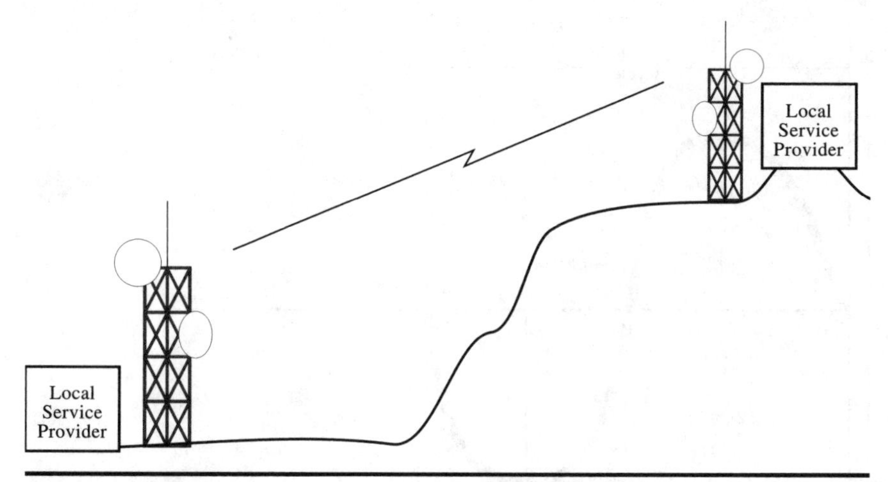

Figure 1.3 Overcoming transmission obstacles such as mountains or valleys.

Figure 1.4 shows a new application for radio-based systems: as a means of bypassing the normal cable route from the local supplier. The dotted line represents how a leased line facility would run from customer location A through the local supplier to customer location B. These leased lines would bear monthly or annual rental costs from the local supplier. Customers could avoid the high rental costs by using a private (owned or leased) radio system.

The use of radio waves in free space requires the conversion of the human voice (or other form of information) from sound to an electrical equivalent. The equipment components used to perform this conver-

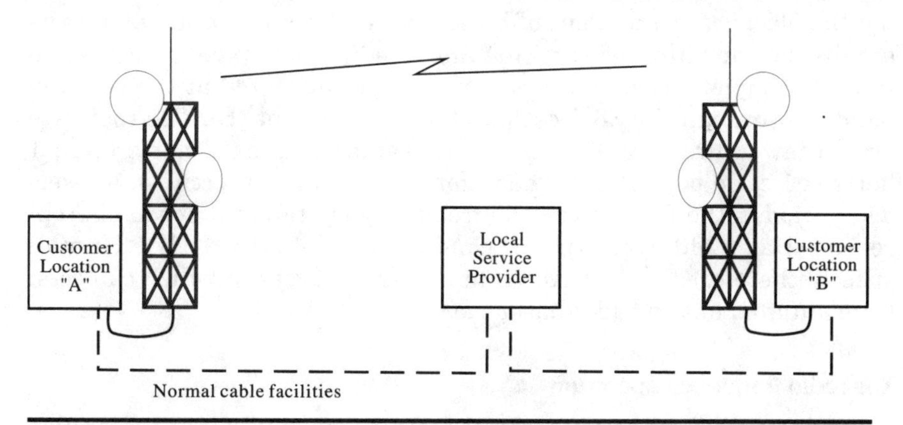

Figure 1.4 Radio-based systems as a means of bypassing the normal cable route from the local supplier.

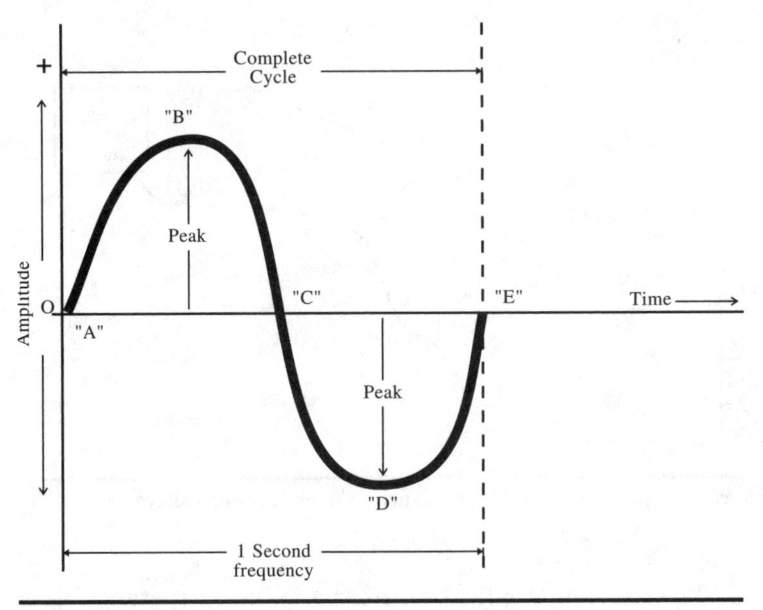

Figure 1.5 The sinusoidal waveform represents a complete cycle in a period of time. In this case, 1 cycle per second equals 1 Hz.

sion operate similarly in every radio system. Although different names may be used to describe the components, it is how they function that is important.

Sound has two constantly changing variables: amplitude (the height of the signal) and frequency (the variable rate of change in a specific period of time). The pattern is normally represented by a sinusoidal waveform, as shown in Fig. 1.5. The waveform representing the electrical equivalent of human speech is a function of both the amplitude (or value of current) and the frequency with respect to time. A complete cycle of the waveform is represented at the starting point (A) through the 360° cycle to the ending point (E). The complete cycle that occurs in a 1-second time frame is called a hertz (Hz); 1 hertz is 1 cycle per second. The number of cycles that occur in a 1-second period is the frequency. The frequency of standard speech is represented at 3000 cycles per second, or 3 kilohertz (KHz). Human speech, therefore, is converted into a 3 KHz waveform which can then be modulated onto a radio-based carrier.

The radio frequency spectrum

Since the waveform represents the electrical equivalent for analog transmission, it will also have a certain velocity with respect to dis-

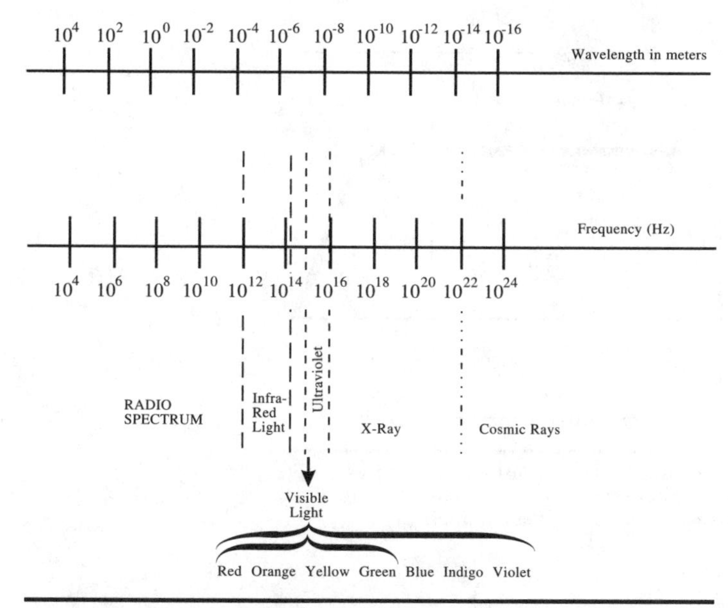

Figure 1.6 The frequency spectrum for radio, light, X rays, and cosmic rays represented in Hz and wavelength in meters.

tance and time. In free-space radio transmission, the electromagnetic wave moves through the air at the speed of light (186,000 miles per second). Radio waves can be produced and transmitted across a wide range of frequencies, starting at about 10 KHz up through the millions of hertz (megahertz, or MHz) and even billions of hertz (gigahertz, or GHz). Figure 1.6 shows the typical frequency spectrum for the transmission of various forms of energy. The spectrum ranges from radio frequencies through light frequencies and finally to X rays and cosmic rays. In Table 1.1 the frequencies shown for radio-based systems are based on a band (range).

TABLE 1.1 Range of Frequencies and the Associated Band Classification

Frequency	Band
<30 KHz	Very low frequency (VLF)
30–300 KHz	Low frequency (LF)
300 KHz–3 MHz	Medium frequency (MF)
3 MHz–30 MHz	High frequency (HF)
30 MHz–300 MHz	Very high frequency (VHF)
300 MHz–3 GHz	Ultra high frequency (UHF)
3 GHz–30 GHz	Super high frequency (SHF)
>30 GHz	Extremely high frequency (EHF)

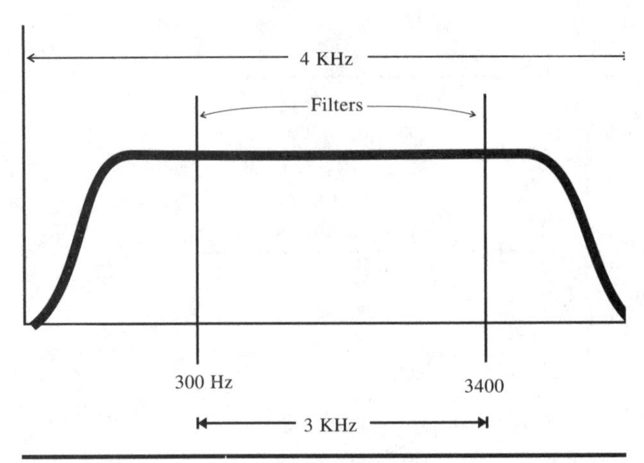

Figure 1.7 The spectrum is divided into 4 KHz slices. The bandpass filters allow conversation to pass at only 300–3400 Hz. This is the band-limited channel.

The modulation process

Once a determination has been made as to what frequency will transmit the information, the information must be modulated (applied) to a carrier that operates in this frequency range. Human speech, as the example being used, generates a range of frequencies of approximately 100–5000 Hz. However, most of the usable and understandable information from human speech is contained within the 3000 Hz range. Therefore, as an economic expedient, commercial transmission systems are manufactured to deliver a range of frequencies (bandwidth) limited to the 3 KHz range. This is accomplished by breaking down the frequency spectrum into 4 KHz slices, whereby the electrical wave is applied to the carrier wave. However, bandpass filters are installed to transmit only those frequencies in the 300–3400 Hz range. In Fig. 1.7 the bandwidth is broken down into the 4 KHz channel, with the bandpass filters installed to allow only 3 KHz of frequency use.

Since 10 KHz is the usual starting point of radio transmission, it would be unreasonable to try to transmit speech information below 10 KHz in the radio spectrum. Therefore, a higher frequency range is used, whereby speech can be modulated onto a carrier signal at the transmitting end. As the speech is introduced to a piece of equipment—in this case, a modulator—it is applied onto a carrier frequency to change the actual wave by producing an envelope of the information. The modulated wave represents the sum of the carrier wave plus the signal to be transmitted. Figure 1.8 shows an amplitude modulation process in which an unmodulated (constant) carrier sig-

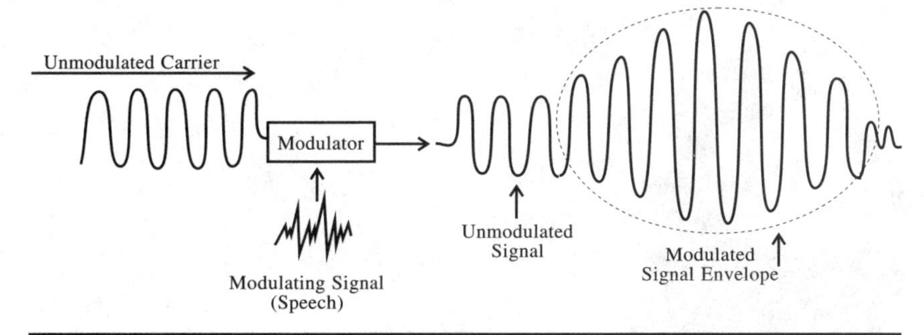

Figure 1.8 As the unmodulated signal is mixed with the modulating signal, the wave changes. The envelope of the modulated signal is the sum of the two waveforms transmitted.

nal is modulated to yield the resultant transmission. Other forms of modulation exist (such as frequency, pulse, and phase) but will not be demonstrated here.

Since the signal is carried in free space (the airwaves), the following factors must be considered in relation to the type of radio transmission system used.

1. A decision must be made on whether to use line-of-sight, point-to-point, or omnidirectional transference via broadcast communications.
2. Noise will always be a factor, since its presence will degrade the signal.
3. Power output will directly affect the distance that the signal will travel.
4. Loss or attenuation will be a factor, since a radio signal will be diminished as it passes through certain insulating materials and will experience gain when it passes through conductive materials or is reflected off other objects.
5. Heavy rain or snow will absorb some of the transmitted signal in certain frequency bands (i.e., microwave and satellite).

Although the above list is not all-inclusive, the most prominent factors are fairly generic for most radio-based systems. Each factor must be given serious consideration prior to setting expectations for performance.

Radio propagation

Depending on the band selected, the characteristics of propagation will vary. In general, when the signal is transmitted through an antenna device, the signal will travel along the earth's curvature (see

Figure 1.9 The transmitted energy follows the curvature of the earth.

Fig. 1.9). As the signal emanates in all directions (or in a point-to-point direction), the energy follows the earth's curvature. In some cases, reflected power off the earth's surface helps achieve the desired result. At lower frequencies (very low, low, and medium bands), the signal follows the curve of the earth's surface in what is typically called a ground wave. The distance that the wave travels is a function of the amount of power generated by the transmitting device. Power output is selected to cover specific distances and areas.

At the high frequency (HF) band, the ground wave is absorbed and attenuated very quickly. However, the radiated energy also has an

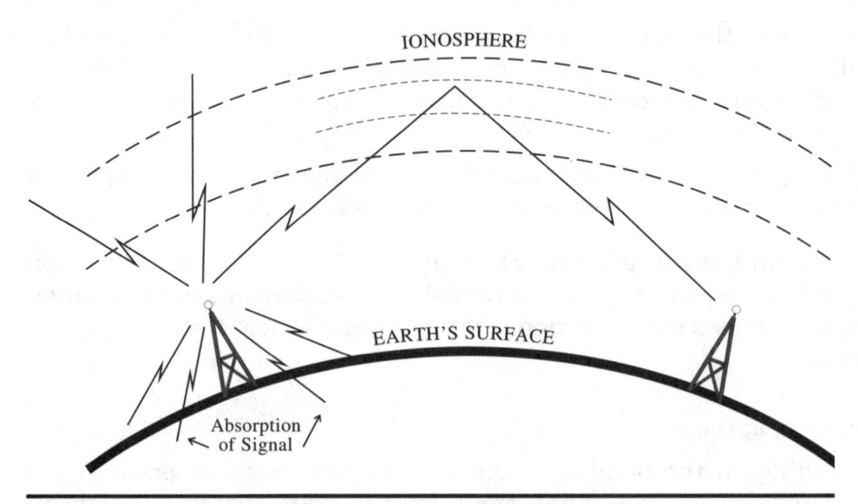

Figure 1.10 In HF transmission, the radio waves are refracted back down to the earth by the ionosphere. Lower power is needed for this system.

upward movement in which the signal reaches approximately 40–300 miles above the earth, entering the ionosphere. In the ionosphere the radio waves are basically refracted at various angles and bounced back down to the earth. This type of transmission enables radio signals to be directionalized and transmitted at much lower power output. Figure 1.10 illustrates this form of high-frequency transmission.

At the very high frequency (VHF) band, the signal is transmitted in straight lines. A directional antenna can be used to direct the signal in a line-of-sight (LOS) path. A certain portion of the signal can be reflected off the ground along this same path and get to the same point. The design of this type of transmission requires great care, because the reflected wave can cause interference. Since the signal is being reflected off the earth's surface (or other surface), the path is longer. Therefore, the reflected signal can arrive later than the direct line-of-sight signal. This delay, or out-of-phase signal, can distort the transmission. Fig. 1.11 illustrates VHF transmission using a line-of-sight signal. The reflected wave is also shown. In order to achieve line-of-sight transmission, antenna height is critical. The greater the distance, the higher the antenna.

At the ultra high frequency (UHF) band, the use of microwave signals is more prominent. In the microwave systems of today, high-range frequencies are used for point-to-point communications. Several channels of communication can be multiplexed together and transmitted across the carrier. These microwave systems are used extensively by telcos, PTTs, and private organizations to carry telephone calls. Two sets of frequencies are required in microwave systems: a transmit frequency and a receive frequency. The lower bands

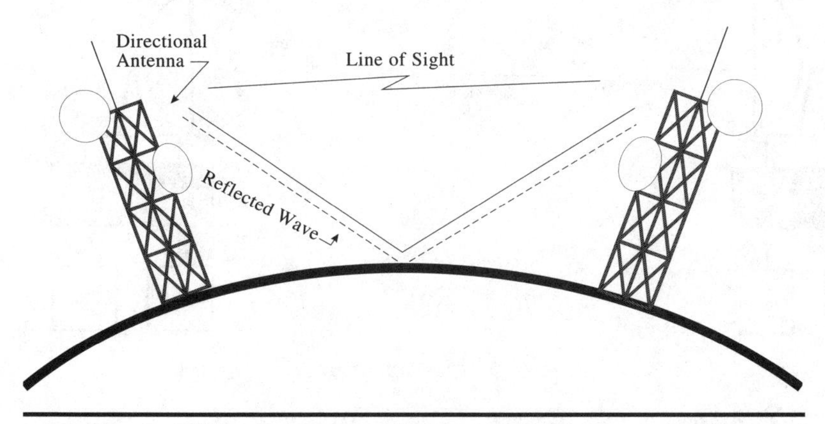

Figure 1.11 In VHF transmission, a line-of-sight path is used. Reflected waves off the earth's surface can cause interference.

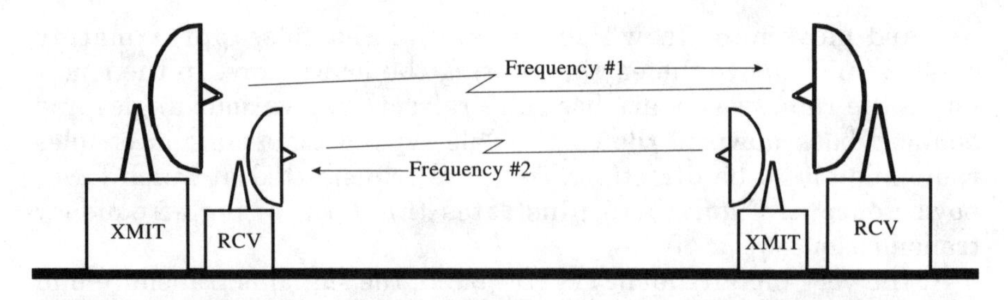

Figure 1.12 Transmission via microwave for telephone systems requires two separate frequencies and two systems.

of the frequency spectrum (LF and HF) are used as a one-way alternating transmission on a single frequency requiring one party to listen while the other speaks. A push-to-talk transmission is used. If both parties try to send at the same time, a jamming effect will render the transmission useless. Very specific transmission protocols are required so that the radio can be used effectively.

In a telephone call, the transmission of voice calls across a radio signal is different. The dynamics of a voice telephone call require simultaneous transmission in both directions since the protocols or rules of telephony are far less stringent. To avoid the jamming effect, two separate frequencies are required. Figure 1.12 shows a microwave transmission using two separate systems (a transmitter and a receiver). Each of the systems has its own frequency. Frequency 1 is used to transmit from west to east, whereas frequency 2 is used to transmit from east to west. Newer microwave systems

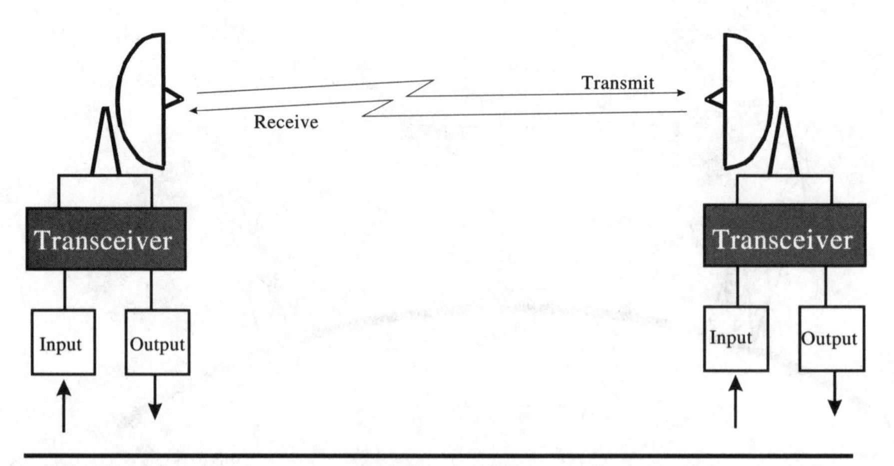

Figure 1.13 Newer systems accommodate telephone calls by using a single transceiver.

combine these transmit and receive functions into a single device called a transceiver, which mixes the input-output onto a single radio system for two-way communication. Figure 1.13 illustrates the newer transceiver system.

A two-way radio telephone system (such as Advanced Mobile Phone Service [AMPS] or Improved Mobile Telephone Service [IMTS]) and a cellular telephone system operate along the same lines as the microwave radio system. One difference is that radio telephony and cellular systems use the lower frequency spectrum, so that distance limitations and output power needed for transmitting apply. However, these systems now have dual frequencies to connect to a radio telephone transmission system. Figure 1.14 shows this general concept for a two-way radio telephone system. Cellular phones work similarly but use different terminology. The base station or cell site (depending on the system used) transmits at much higher wattage output than the mobile set (or cellular phone). Output at a base station or cell site can be in the 15–30 watt range, whereas the mobile set will output at 5 watts for older systems, 3 watts for cellular, and .3 to .6 watts for portable cellular sets.

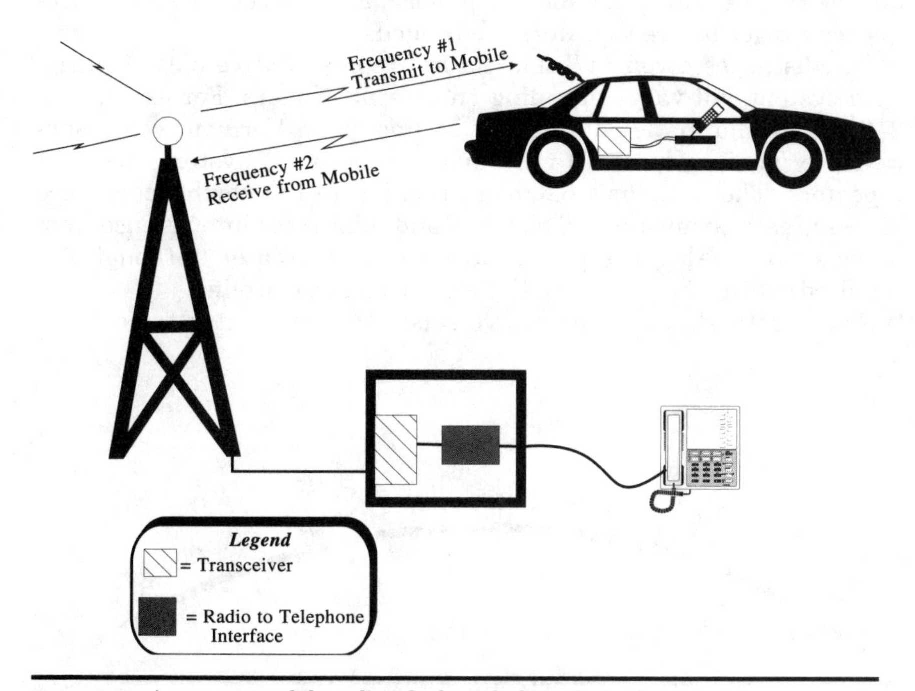

Figure 1.14 A two-way mobile radio telephone link works in the same manner as the higher-end radio systems.

TABLE 1.2 Typical Microwave Distances Before Repeaters Are Necessary

Frequency	Approximate distance
2–6 GHz	30 miles
10–12 GHz	20 miles
18 GHz	7 miles
23 GHz	5 miles

Microwave repeater systems

As radio-based systems evolved, it was discovered that a microwave transmission could cover greater distance through the use of repeaters. Basically, a repeater receives the radio signal on one antenna, converts the signal back into its electrical properties, then retransmits the signal out across a new transmitter. In some cases, the repeater changes the signal from one frequency to another. Line-of-sight transmission is used, and the height of the antenna is important to maintain this capability. However, limitations from power, path, and other sources still restrict the distances covered. The distance limitations vary for microwave transmissions in various frequency ranges. Table 1.2 shows the normal distances for several frequency ranges before repeaters are needed.

The distances given in Table 1.2 are representative only. Clearly, each system will vary depending on external factors. For example, a 2–6 GHz radio installed atop Mt. Killington in Vermont could successfully achieve a 50–70 mile distance between systems without repeaters. The antenna's height in relation to the earth's curvature allows greater transmission distance and minimizes interference from other sources. Also, the power output can be greater (although not required) at this height, since there's nothing else around.

Figure 1.15 shows a microwave repeater system at a typical dis-

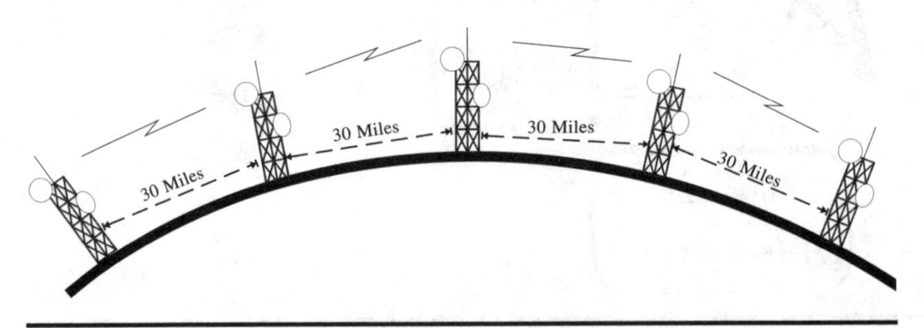

Figure 1.15 Repeaters are used to generate greater distances on microwave transmission.

tance of 30 miles. The system is for illustrative purposes only, since other considerations would affect transmission. In the telephone systems across the world, this type of transmission has been used extensively. It is still widely used in networks across great distances.

Satellite radio communications

As telephone systems continued to evolve and the need to transmit information over greater distances grew, a newer radio-based system emerged. In 1960 microwave radio signals were transmitted up into the atmosphere to a repeater floating in space. Called *satellite communications,* the system was originally designed to bounce radio waves off an artificial object that was orbiting the earth. Earlier attempts to bounce radio waves off the moon were not as successful as had been hoped. The system worked, but the returning signal was so weak it couldn't be used. A similar attempt to bounce radio waves off an inflated weather balloon met with equally unrewarding results. Therefore, an active rather than a passive system became the next logical step.

An orbiting satellite offers several distinct advantages. A stronger signal can be obtained, and over very long distances the transmission signal requires only a single repeater (i.e., the satellite). The satellite can be located in a polar, inclined, or equatorial orbit. The orbit can be either circular (equidistant) or elliptical (nonconcurrent) at different heights above the earth's surface. The early satellites were launched into elliptical orbits at lower heights above the earth than satellites of today. The orbit around the earth took from 1 to 2 hours, depending on the height and path. Therefore, the earth station equipment had to track the satellite and could transmit only for limited periods when the satellite was visible. The system proved impractical for commercial use, since several satellites were needed to provide constant communication and the moving antenna equipment required the constant reaiming of dishes. Therefore, a circular orbit around the equator (equatorial orbit) at a height of 22,300 miles was selected. A satellite at this height takes 24 hours to orbit the earth, resulting in what looks like a stationary object. In fact, this is called a *geostationary* or *geosynchronous orbit.*

Further, at this height a wide footprint (area of coverage) of the radio beam (at 170° dispersion) back on the earth's surface produces a very large area of coverage—approximately a third of the earth's surface—from a single satellite. Thus, only three satellites are needed to provide almost total coverage.

Once the position of the satellite and the coverage areas are determined, the rest is straightforward. A microwave transceiver is used to

TABLE 1.3 Frequency Ranges for Satellite Transmission

Band	Transmit (Uplink)	Receive (Downlink)
C	5.925–6.425 GHz	3.7–4.2 GHz
K_u	14.0–14.5 GHz	11.7–12.2 GHz
K_a	27.5–31.0 GHz	17.7–21.2 GHz

transmit the signal up to the satellite. Transmissions are in the super high frequency (SHF) range. Table 1.3 shows the frequency ranges for both transmit (uplink) and receive (downlink). Other frequencies are being considered for use in radio transmission systems, but these have not yet been codified.

The low end of this spectrum, the C band, is shared in commercial microwave systems. Therefore, close coordination must exist in the frequency allocation, spacing, and antenna location to prevent interference. However, as with all radio transmissions, the lower end of the spectrum is more desirable because it is less susceptible to attenuation and absorption (from rain and snow). The higher the frequency used, the greater the risk of absorption. Once an appropriate system is chosen, uplinks are used to transmit to the satellite. Fig. 1.16 shows the

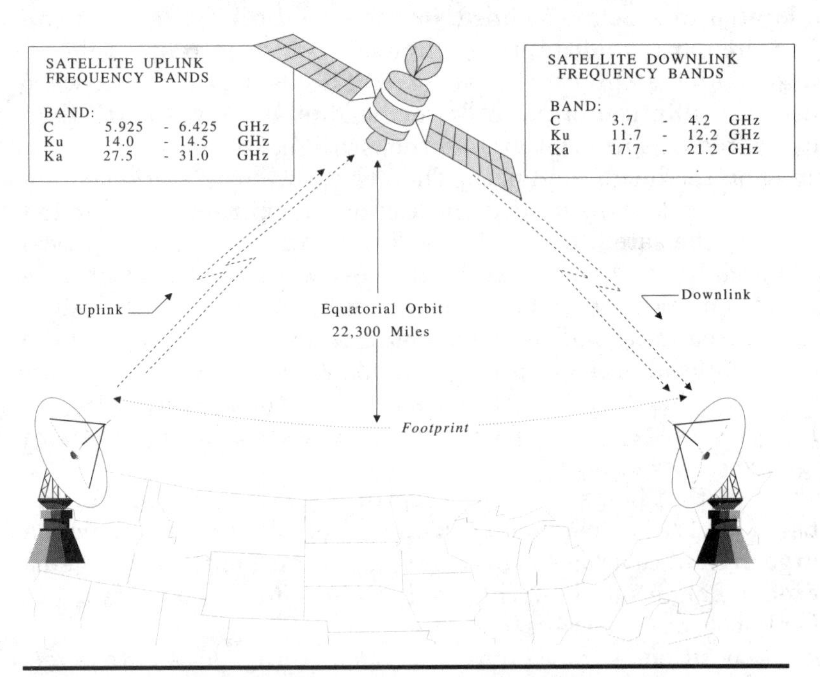

Figure 1.16 Satellite transmission covers a very large footprint.

satellite and its associated footprint. A single satellite can embrace North and South America in its radio beam path.

A terrestrial link (microwave, copper, coaxial, or fiber) is run from the customer's location to the nearest uplink for transmission and reception of satellite communications. Figure 1.17 shows a U.S. connection linked through various means. Internationally, the arrangement will differ, depending on the degree of access allowed in a particular country and the rules of its post telephone and telegraph (PTT) organizations.

Satellite transmission has been used extensively for long-haul communications and for transoceanic international communications. However, the costs associated with satellite launch, electronic components, earth station equipment, and orbit life have all been somewhat detrimental in the overall picture. Whereas other technologies (such as microwave and fiber optics) offer declining scales of cost, satellite costs have remained high. For this reason, many industry users opted to use this transmission system selectively. Domestic use of these systems was expected to fade quickly into the past.

Chapters 2 and 9 examine the reemergence of satellite-based systems as new applications appear on the horizon. However, steady installations *have* taken place in developing countries around the world as a quick means of providing national and international connectivity. Countries that lack the infrastructure of terrestrial-based

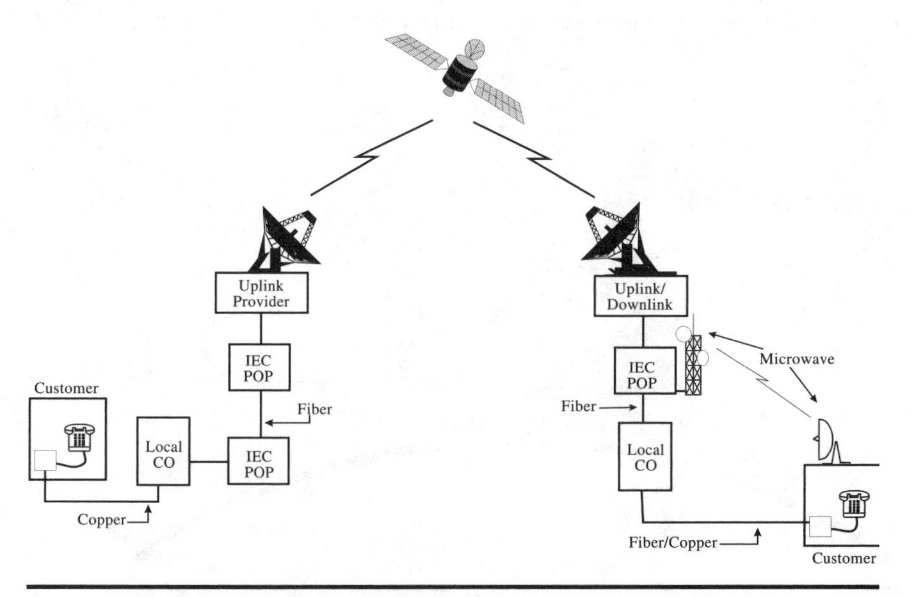

Figure 1.17 Various means are used to get to the uplink and downlink.

systems are ripe for deployment of radio-based systems. Compared with a copper-based wired environment, the wireless satellite alternative certainly offers a better cost-to-benefit ratio and return-on-investment picture.

A large amount of bandwidth is available with satellite radio systems, since the frequency spectrum can be allocated on a fixed-access or demand-access basis. Suffice it to say that the capacity is available if the user wishes to experiment or use the system full time. Since the signal must be transmitted up 22,300 miles to the satellite, the rental cost of a channel (or other portion of the bandwidth) is insensitive to distance. For now, it is safe to assume that the benefits outweigh the transmission disadvantages; therefore, the satellite will be around for a long time to come.

Other transmission systems

Another form of wireless transmission is troposcatter radio, which has primarily been used by the military (starting in the 1960s and continuing to today). The troposcatter system transmits in the HF/VHF range and broadcasts the signal up through the earth's atmosphere to the troposphere. At this layer, the signal is refracted back down to the earth's surface. The advantages of this system include very long distance transmission paths, lower power output requirements, and highly directionalized communications. In the higher frequency ranges, very deterministic paths of communication can be plotted. Troposcatter radio has not received acceptance in the

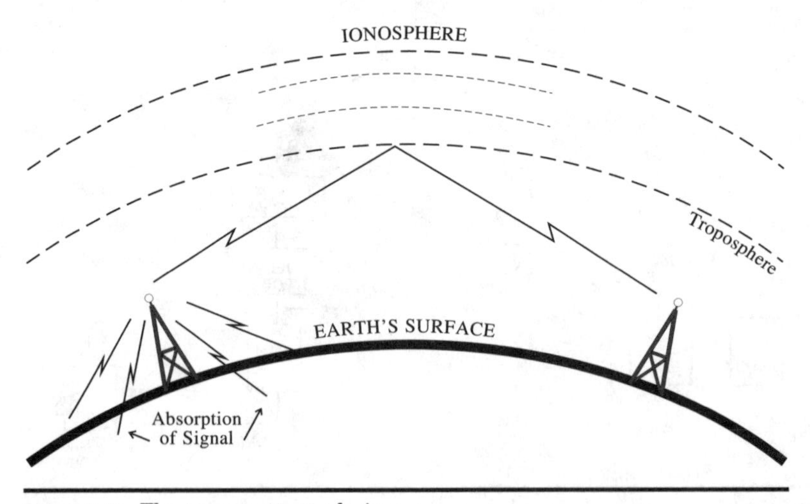

Figure 1.18 The troposcatter technique.

commercial world (carriers and customers). Figure 1.18 illustrates the troposcatter technique.

Meteor burst communications is another limited application for radio transmission systems. A radio wave is transmitted along a predefined path up through the earth's atmosphere to space. Very definitive paths are selected to transmit the information so that the signal will come into the direct path of free-space meteors. Upon entering the meteor path, the radio wave is bounced back at a specific angle to the earth's surface, where it is then delivered to a receiver. Bursts of data are blocked into a data grouping for transmission and then burst out over the spectrum. This system too is primarily a military transmission. Some limited commercial applications have emerged, but most commercial users are reluctant to experiment with the service. Telemetry, weather, and navigational systems are the primary applications. However, any block-mode data transmission can be used in the meteor burst spectrum.

Light-Based Systems

Another wireless transmission system that is both reliable and inexpensive has proved to be less popular than radio-based systems. Infrared light has long been used as a transmission medium for short-haul local communications. While the benefits of radio are many, the quality impairments, limited frequency spectrum, and rain attenuation problems all must be considered equally as disadvantages. The primary advantages of infrared systems are that they are easily transportable, typically can be set up and operating within an hour, and do not require expensive licenses or right-of-way permits. For all these pluses, the main disadvantages are the limited distance and limited bandwidth. In addition, infrared transmissions are subject to even more fading and absorption than microwave systems.

A pinpoint-beam generator in the infrared (invisible) light spectrum operates at a frequency range of 10^{12}–10^{15} and a wavelength of 10^{-4}–10^{-7} to produce high-speed communications. What should also be understood is that in this context infrared is carried in open airwaves. If a different medium, such as very pure glass (fiber) were used, the spectrum of the light would be in the infrared range. It is the fiber that achieves the high-speed, reliable transmission, since the impairments of the air are overcome on the glass.

Most consumers are familiar with wireless infrared transmission, although they may not realize it. The remote controllers for TV sets, radios, lights, and other appliances use infrared signaling. Clearly, these low-powered controllers are very limited applications, but they have all the properties of infrared transmission. They can be used only for short distances and are highly directional (point to point), invisible light transmitters.

Also, these systems are one-way devices; separate transmitters are needed for two-way communications. Most providers recommend distances of less than 1 mile to maintain the reliability and integrity of infrared transmission.

Chapter

2

Radio-Based Systems

Troposcatter Systems

As we have seen, in addition to typical line-of-sight systems like microwave and satellite, systems not based on a line of sight are available. In general, these are an extension of the microwave communications service. However, beyond their normal microwave characteristics, troposcatter systems allow for greater transmission distances. The distances allowed with microwave systems vary from a low end of 5 miles to a high end of approximately 30 miles. If there is a need to provide communications over longer distances, back-to-back (or repeatered) systems are used. (See Fig. 1.15 in Chap. 1.) Price and the real estate issues notwithstanding, the number of systems increases exponentially as the distance increases.

To overcome this predicament, bouncing radio signals off the clouds (in actuality, the troposphere of the earth's atmosphere) at an angle and refracting them back to the earth adds some dimension to meeting the needs of the user, as shown in Fig. 2.1. In this case, the bounced radio signals can increase the distances covered and reduce the number of hops needed to complete a radio shot. The distances allowed with troposcatter systems approximate upwards of 400 miles. They have been used extensively in the military for voice, data, and telemetry communications over great distances where the number of repeater sites would make the cost of the required signal too expensive. By increasing the distance of the signal transmission path and decreasing the number of hops needed, the military was able to install systems that carried from 12 to 240 voice communication channels. Clearly, these systems afforded certain advantages, particularly in light of the tactical nature of some military communications. Further uses have been found in retail industries, utility companies,

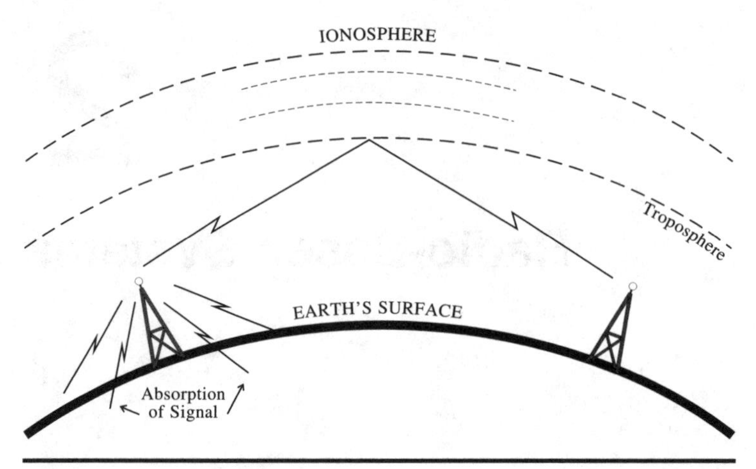

Figure 2.1 The use of troposcatter bounces radio waves off the clouds.

long-haul pipeline companies, and long-haul communications systems for government agencies (such as forestry services and agriculture).

How the system works

As noted in Chap. 1, in the early days of satellite transmission scientists and engineers attempted to bounce radio waves off floating satellites, weather balloons, and even the moon. In a similar fashion, in troposcatter transmission the radio signal is bounced off the troposphere of the earth's atmosphere. As the signal is directed toward outer space, a certain amount of the radio energy is refracted (reflected back) toward the earth to a destined receiver station. With its varying degrees of density, the troposphere acts in a manner similar to glass in fiber-optic transmission. When the density (refractive index) of the glass is varied, the light is bounced back into the path of the glass toward the distant end. Similarly, the varying densities and indices of the earth's atmosphere can direct the radio beam to a specific point, depending on the angle of refractance. Some consistencies must be applied with this type of transmission, since the angle of refractance should be kept at a minimum and should be congruent with the receiver angle. The transmitter will form a scatter angle whereby the signal can be propagated back in appropriate form. This angle will also have some common volume between the transmitter and the receiver.

Advantages of troposcatter

The key advantages of a troposcatter transmission system include:

- Reduced number of hops to cover greater distances

- Highly reliable transmission across large voids such as bodies of water, valleys, and mountain ranges
- Ability to reach remote areas with limited amounts of equipment, such as long-haul pipeline services
- A mix of services for low-speed voice data—typically 2400–4800 bits per second (bps), although in some cases 9600 bps have been achieved
- Tactical advantages when used in military applications for long-haul communications systems requiring multichannel services
- Highly reliable telemetry and tracking systems
- Reduced real estate requirements for system shots in the microwave frequency range (eliminating the need for repeaters every 20 to 30 miles)
- Ability to transgress various border facilities and boundaries where taxation and tariffs are extraordinarily high

Disadvantages of troposcatter

The key disadvantages of troposcatter transmission include:

- Expensive equipment requiring high up-front costs
- Fading of the signal, although the problem can be somewhat predictable
- Limited channel capacity compared with other systems
- Fallback of speeds on systems where turbulence in the atmosphere leads to fading or deterioration of the signal

Operating bands for troposcatter systems

Several bands are allocated for the most efficient use of troposcatter systems. These are shown in Table 2.1. The allocation of these bands has worldwide significance because of the distances covered, the spectrum used, and the fact that the signal may traverse space over other

TABLE 2.1 Bands Allocated for Troposcatter Systems

Low band	High band
350 MHz	450 MHz
755 MHz	985 MHz
1.7 GHz	2.4 GHz
4.4 GHz	5.0 GHz

regulatory agencies and governments. Therefore, the consequences are an issue not only for the Federal Communications Commission (FCC) but also for the International Frequency Regulatory Board (IFRB), which coordinate use of the radio spectrum worldwide.

These frequency ranges are also used in other transmission systems, including the two-way mobile FM radio system in the 350–450 MHz range, the cellular and industrial scientific and medical (ISM) bands in the 755–985 MHz range, and the microwave and global services mobile (GSM) standard, now the personal communications services range, in the 1.7–2.4 GHz range. This is merely a comparison of the various services, since the spectrum is allocated according to need and the frequency clearances are assigned by the FCC and the IFRB.

System configuration

Troposcatter systems use the same basic elements as other radio systems for the transmission and reception of frequency signals. The major differences in these systems are the directionality and the delivery of the signal from the transmitter to the receiver and vice versa. No two systems are alike; there will always be slight differences when dealing with a radio-based system. However, the underlying concept is still the same: An electrical current (whether a digital or an analog input is used) is delivered to the radio transmitter and is then modulated onto a radio carrier wave. From there the wave is transmitted through the air to the receiving station. At the receiving station, the radio wave is demodulated from the carrier wave back to the unmodulated state, extracting the information and delivering it to the electrical output. The electrical current, identical to the input, is then delivered from the output to the recipient of the information, whether it is a voice, telemetry, or data terminal device. Many different configurations can be used in the equipment, but this simplified description sums up the process nicely.

Other characteristics of troposcatter systems include the sensitivities of the equipment, the risk of overscatter and the ensuing loss effects, the use of multiple antennas and receivers, and the significance of the angles for transmission and reception. These considerations are beyond the scope of this discussion but are factors that should be researched if a troposcatter system is considered.

Capacities

Troposcatter systems offer some reasonable voice grade channel capacities, but they are limited compared with some of the other radio-based systems already discussed. The obvious application for a voice grade channel assignment using a troposcatter system is the ability to over-

TABLE 2.2 Channel Capacities Based on the Frequency Band Used

Band	Channel capacity
350–450 MHz	60, 120, 180, or 240
775–985 MHz	60, 120, 180, or 240
1.7–2.4 GHz	12, 24, 36, 48, 72, or 132
4.4–5.0 GHz	24 or 36

come the distance limitations of line-of-sight services. However, this application seems best suited to tactical services, such as the military operation of troposcatter systems. The spreading of the bandwidth over these radio frequencies requires some form of diversity when transmitting. This leads to a somewhat inefficient use of the radio frequency spectrum, but it is the only way to ensure that the system delivers the information. Thus, two forms of diversity are normally used: space diversity and frequency diversity. Table 2.2 summarizes the changes.

Microwave Radio

The primary goals for communicating between or among radio systems are basically the same regardless of the method used. These goals include:

- High-quality transmission that offers a range of services such as voice, data, video, facsimile, and LAN interconnection at an acceptable bit-error rate
- Exceptional reliability offering maximum availability of the services regardless of the transmission
- Time and transport systems that offer low-cost transmission services and a quick return on investment
- Highly flexible services that offer a range of interfaces to accommodate any type of transmission and the ability to superrate or subrate the capacity depending on the transmission need
- Ease of integration into other network transmission services such as local telephone companies (or PTTs), long-distance carriers, international gateways, and private networks

Evolution of microwave

Radio wave transmission through the atmosphere was introduced in the 1900s after Guglielmo Marconi demonstrated the ability to transmit signals on a global basis using high-frequency transmission in the 3–30 MHz band. Through several improvements and modifications, developers introduced the ability to transmit at much higher frequencies. In the early 1930s the frequency bands from 30 MHz to 12 GHz were conceptualized and implemented. These higher frequencies proved to offer several advantages over existing high-frequency radio systems. These advantages include, but are not limited to, the following.

1. The atmospheric noise that affects and degrades a high-frequency transmission decreases with the increase in operating frequency transmission. The decrease in noise goes below what is known as the thermal noise threshold. With the reduction in atmospheric noise, the radio transmission through the air at 100 MHz and above begins to match or exceed the operating characteristics of the copper-based circuits used in the telecommunications industry.
2. The propagation of the signal occurs in the troposphere (the earth's lower atmosphere), providing more stability than the ionospheric propagation used in high-frequency transmission. There are also potential pitfalls, since tropospheric propagation is affected by weather conditions such as air pressure, ambient temperature, and turbulence.
3. As the operating frequency increases, the propagation of the radio wave begins to emulate an optical or visual straight line in the path. This line of sight allows the reuse of frequencies in close proximity so long as the multiple systems using the same frequency are not in the line of sight to each other. This reuse of frequencies is of some benefit in a limited-frequency spectrum.
4. Capacities using higher frequencies exceed the current spectrum in the HF band by at least a thousandfold.
5. The bandwidth which the transmitted information is modulated onto a thousandfold wider than the HF band. Therefore, far more resiliency and flexibility can be achieved. A microwave transmission can be used to modulate and transmit a more robust array of services from multichannel voice frequency transmission to broadband communications, such as video and high-speed data services.
6. The cost per circuit mile is very attractive for higher-frequency transmissions, since the equivalent analog or digital telecommunications channel is priced on far more costly components.

Frequency bands allocated

The frequency spectrum of microwave ranges from 1 GHz to 30 GHz. These ranges, shown in Fig. 1.5 in Chap. 1, are presented in more detail in Fig. 2.2. The principal advantage of the microwave spectrum

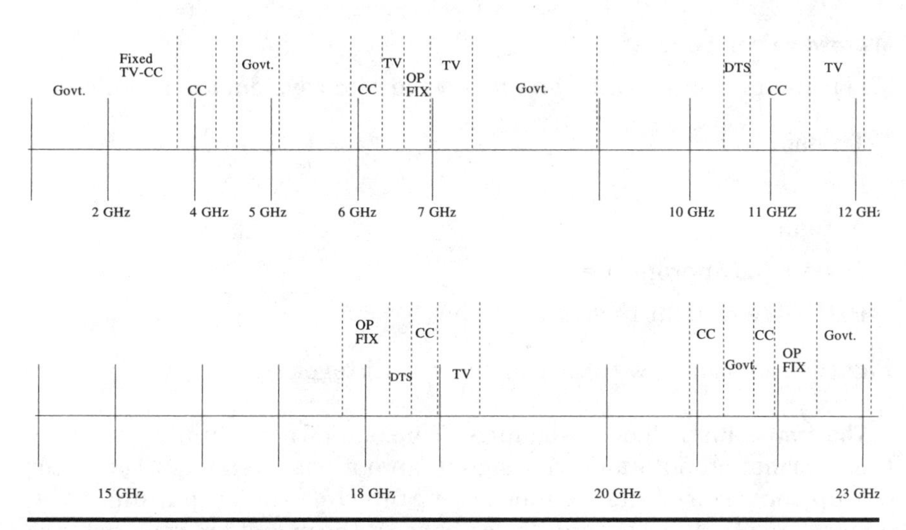

Figure 2.2 Frequency spectrum of microwave.

is the ability to get around physical obstructions along cable rights-of-way. The capacity of microwave is another distinct advantage. In 30 MHz of bandwidth, a radio signal can be subchannelized to carry 6000 voice channels. Several operating systems and frequencies are assigned to various agencies, including:

governments
common carriers (telephone companies and long-distance carriers)
private operating systems (utilities and commercial businesses)
TV operations for transmission of television signals

The competition for this limited bandwidth is intense. Looking at the spectrum and its limitations, one can see reality setting in. The industry is running out of radio frequency (RF) spectrum. An example is a move by the Federal Communications Commission to reallocate some of the RF spectrum in the 1.7–2.3 GHz range for the newer offerings of personal communications services (see Chap. 5). However, in defense of this move by the FCC, it is fair to say that some microwave operators have been idly sitting on RF bandwidth for years, leaving a strained capacity and denying other valid uses to organizations and agencies that have a true need. Since the FCC is chartered to administer this frequency spectrum in the United States, the onus is on the agency to seek out bandwidth "hogs" that have licenses to operate but hold the spectrum in reserve for some future application that may never develop.

Microwave components

The primary components of a microwave radio system are as follows:

Transmitter

Receiver

Antenna

Towers (as appropriate)

Path (line of sight through the air)

Figure 2.3 shows how these components fit together.

The transmitter. The transmitter consists of a modulator to apply the information onto a carrier signal, an optional frequency converter (an up converter shifts frequencies up) to shift the output-input frequency, an amplifier output device to boost the signal for long-distance transmission, and an antenna coupler with a circulator to combine or separate the received signal from the transmitted signal so that both can use the same antenna. Figure 2.4 illustrates these components.

The receiver. The receiver consists of another circulator to separate the received signal from the transmitted signal on the same antenna, a frequency converter to shift the frequency down, and a demodulator unit to recover the information from the carrier wave. The receiver is diagrammed in Fig. 2.5.

The antenna. Once the frequency band and radio system have been selected, the correct antenna must be used to provide optimal performance, ensure high reliability of the transmission, ease the frequency

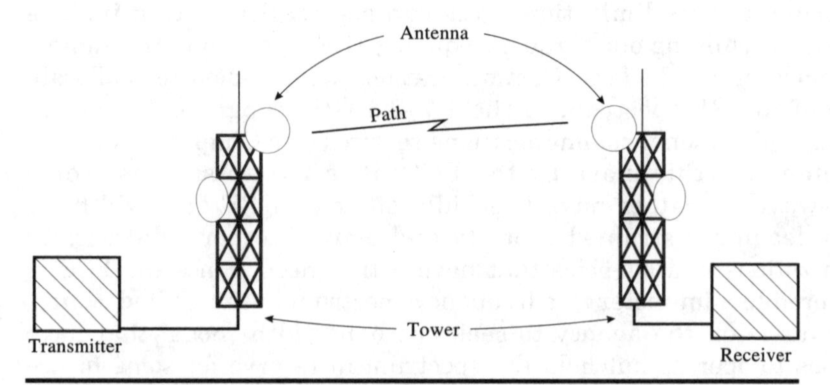

Figure 2.3 The basic components of microwave radio systems.

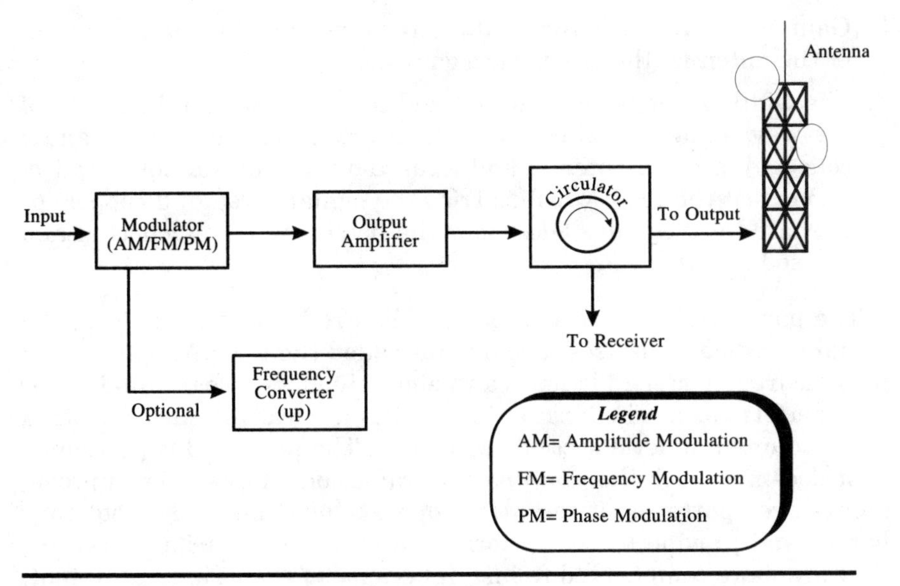

Figure 2.4 The basic components of a transmitter.

selection process, and provide for regulatory compliance. The antenna is a critical link between the transmitter and the receiver. For point-to-point transmission, the antenna is highly directional, shaped like a parabola or horn. Some of the considerations in selecting an antenna are:

1. The bandwidth must support the operating frequency with minimal distortion.
2. The antenna must meet regulatory constraints, since size will affect gain, beamwidth, and tower design.

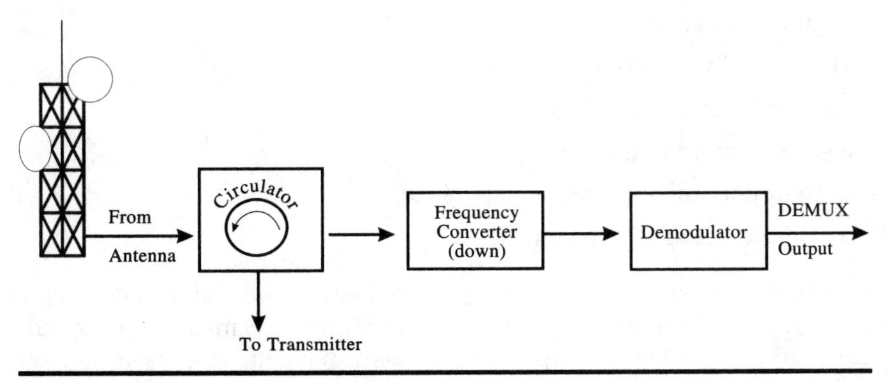

Figure 2.5 The basic components of a receiver.

3. Gain is a power ratio that is directly proportional to size. The larger the antenna, the greater the gain.
4. The width of the beam collected and reflected back to the center of the waveguide will affect overall performance. The radio beam is collected into the antenna and focused onto a central point (collector) called the feed, which carries the signal through a copper- or air-based waveguide. The wider the beam the more collected and focused into the feed.

The parabolic antenna is a curved dish reflector that collects the signal energy and focuses it to a point called the feed. Antennas support all frequency bands and can be classified as either standard or high-performance. The horn antenna uses a parabolic surface with a horn mounted under a section of the dish. The horn feed is positioned so it does not block the main path of radiation. Horns offer improvements over patterns of radiation, operate over multiple frequency bands, and provide better performance than do parabolic. However, they are more complicated to install because of their size and weight. They are also more costly.

A radome antenna is designed for adverse conditions such as heavy wind loads or accumulations of ice, snow, or dirt. Radomes are basically covers that protect the horn feed and dish. Unfortunately, the radome also decreases the antenna gain.

Towers. Towers play a very important role in a microwave system. Rising above the obstacles along the path and accommodating the earth's curvature over distances, towers are used to guarantee line of sight between the systems. Several factors must be considered when choosing a tower:

Cost

Regulatory restrictions

Air traffic in the area

Soil conditions or rooftop conditions

Weather (wind loading)

Self-supporting or guyed construction

Waveguide characteristics

Standard or normal towers are prefabricated and delivered in sections for on-site assembly. Guyed towers are the most economical. The heavier and more expensive self-supporting towers are used when higher antennas are required or when platforms are needed for mounting the antenna. Self-supporting towers may be triangular or square. Triangular towers are preferred for their stability.

The path. The propagation path of a radio signal includes the direct wave, a reflected wave, and a surface wave. These combine to form the ground wave. Another portion of the wave, the refracted wave, is a function of atmospheric conditions. Figure 2.6 shows the path between two antennas with the various components highlighted. Each of these components affects the transmission of the radio signal in terms of loss (attenuation) and distortion. In microwave systems, the direct or free-space wave is the controlling influence. However, the refracted wave must be considered as a deterrent to high-quality transmission, since it can cause multipath fading.

Other considerations must be looked at in path selection. For example, with line-of-sight systems, it is the operator's responsibility to be aware of any construction plans along the path. A new building erected in the middle of the path can render transmission totally useless. Performance issues such as rain absorption must also be considered. Any system over the 8 GHz frequency range is prone to rain attenuation—that is, raindrops scatter or absorb the energy, impeding the reception of a high-quality signal. The higher the frequency, the greater the risk.

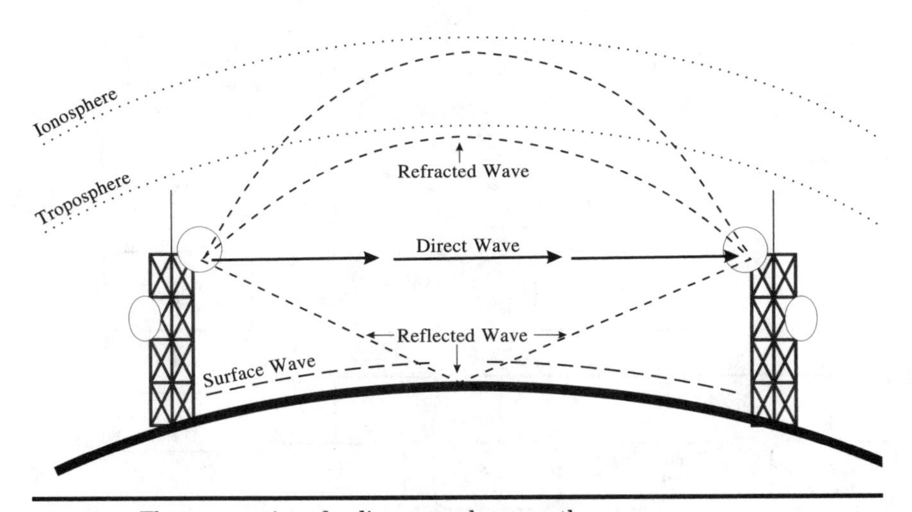

Figure 2.6 The propagation of radio waves along a path.

Licensing and regulation

In the United States, the Code of Federal Regulation (CFR) is the solidification of general and permanent rules published by the federal government. The code is divided into 50 separate titles representing broad areas subject to federal regulation. Each title is divided into chapters which bear the name of the issuing agency. The chapters are subdivided into parts covering specific regulatory areas.

The FCC. The FCC is responsible for the administration of rules and regulations governing the management of the frequency spectrum for domestic, non-federal government users. Federal government users are regulated by the Interdepartment Radio Advisory Council (IRAC), which resides with the National Telecommunications Information Agency (NTIA), a division of the Department of Commerce. Obviously, there are "hand across the agency" arrangements to allow for equality in the administration of the rules for all users. The NTIA's *Manual of Regulations and Procedures for Federal Frequency Management* is the equivalent of the FCC's rules and regulations. The FCC divides its functional work areas among four bureaus. Figure 2.7 shows the organizational flow of the FCC.

The common carrier bureau. The common carrier bureau is responsible for developing, recommending, and administering policies and procedures to regulate services, facilities, rates, and practices of all entities that provide interstate or foreign communications services

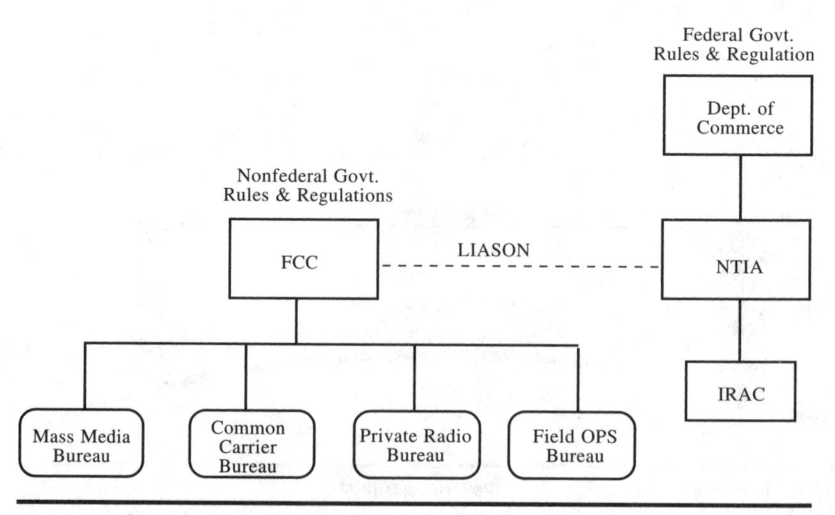

Figure 2.7 Organization of the FCC radio and liaison bureaus.

TABLE 2.3 FCC Part 21 Common Carrier Spectrum

Frequency (GHz)	Bandwidth (MHz)	Analog channels	Digital (Mbps)
2.11–2.18	.8	48	1.544
	1.6	96	3.1
	3.5	288	6.3
			12.6
3.7–4.2	1.6	96	3.1
	20	1500	90
5.9–6.4	4.9	300	12
	14.8	960	90
	29.65	2400	125
10.55–10.68	1.25	—	1.5
	2.5	—	3.1
			4.8
10.70–11.70	20	1500	45
	40	2700	90
			135
17.7–19.7	5	Video	6.3
	10	Video	12.6
	20	Video	25
	40	Video	45
21.225–22.975	25	Video	6.3
	50	Video	25
	100	Video	25
31.0–31.3	25	Video	6.3

for a fee. Included are wire, radio, cable, and satellite services, as well as ancillary services. A specific FCC regulation—Part 21, Domestic Public Fixed Radio Services—prescribes the way that portions of the radio frequency spectrum will be used for domestic common carrier services. Table 2.3 summarizes the FCC bandwidth for each of the Part 21 frequency bands, along with the analog channel and digital capacities that fall within the domain of the FCC bandwidth allocation.

The private radio bureau. The private radio bureau is responsible for developing, recommending, and administering policies and procedures for the deployment and regulation of private radio services. These include national and international radio utilization by individuals, business organizations, state and municipal (local) governments, and others licensed to operate their own communications systems for private use to accomplish their chartered mission statement. The FCC's Part 94, Private Operational Fixed Microwave Service (OFS), delineates the way that operational fixed radio services may

TABLE 2.4 FCC Part 94 Operational Fixed-Service Spectrum

Frequency (GHz)	Bandwidth (MHz)	Capacity	
		Analog channel	Digital
.928–.952	.025	9.6 Kbps	19.2 Kbps
.952–.960	.05	3	—
	.1	6	—
	.2	12	—
1.85–1.99	5	300	12 Mbps
	10	780	25
	10	600 and T1*	
2.13–2.20	.8	48	1.5
	1.6	96	3.1
	1.6	120	3.1
6.525–6.875	.8	48	1.5
	1.6	96	3.1
	1.6	120	3.1
6.530–6.780	5	300	12
	10	780	45
10.55–10.68	1.25	—	1.5
	2.5	—	6.3
17.7–19.7	2		
	5	300	6.3
	6		
	10	600	12.6
	20	1200	45
	40	1200	45
	80	—	45
21.8–23.6	25	960/Video	6.3–12.6
	50	—	45
	100	—	—
31.0–31.3	25	—	6.3 Mbps
	50	—	—

*Note: A T1 is a carrier system that multiplexes 24 digital channels together into a 1.544 Mbps data stream.

be operated and licensed in the microwave frequency spectrum. Table 2.4 shows the FCC bandwidths along with the analog channel and digital capacities in the Part 94 frequency bands.

In order to obtain a license, all radio system users must submit an application and pay certain fees. These fees are applicable for all new services, modifications of existing licenses, and renewals of existing services. Every transmitter must have evidence documenting the authorization to operate. This is called the *station license.* The steps required to obtain a license apply to all permanent licenses, regardless of whether they are Part 21 or Part 94. These steps include:

1. Preengineering—system design, path survey, and initial site layout
2. Frequency selection based on prior coordination
3. FCC construction permit request
4. FAA notification when tall towers are used
5. Detailed engineering—equipment acquisition begins
6. FCC construction permit granted
7. Installation and testing
8. FCC license issued—in-service operation begins
9. Postengineering—submission of as-built drawing

Frequency coordination is required before submitting an application. The requester must verify that the operation of the proposed system will not interfere with that of nearby systems. Several companies provide a service to analyze the frequencies in use through a database scan of all systems and locations in use nearby. They will advise the user of what frequency is available or select a specific frequency and verify that it will work without interference. After the analysis is completed, all other licensees in the area are notified of the results and given 30 days to respond with objections to the new use prior to frequency assignment.

The licensing period can take from 6 months to 2 years, so sufficient lead times are essential. A typical license is issued for a 5-year period, although the FCC has been considering a 1- or 2-year renewal period for all operators.

Costs of microwave systems

Costs of complete microwave radio systems vary from site to site. However, certain one-time cost decisions should be considered prior to going into a system. Table 2.5 summarizes the initial, one-time installation costs. Users are often shocked to find that when a start-up system is estimated at, say, $40,000, the real cost jumps to $80,000 or $100,000 because of additional construction.

Next, the ongoing costs must be added to the one-time system price. All fees and expenses must be included for a true assessment to be made. Table 2.6 provides a format for computing the ongoing costs for a 5-year (or longer) period. The supplier should offer to provide estimates of what the typical numbers will be.

Once the total one-time costs and the 5-year ongoing costs have been computed, they can be reviewed against the equivalent leased-

TABLE 2.5 One-Time Costs for a Microwave System

Item	Cost ($)
Alarms	
Ancillary interface equipment	
Battery system	
Channel bank equipment/multiplexer	
Concrete pad (as needed)	
Cover, radome, or shroud	
Electrical wiring	
Engineering/architectural fees	
Fencing (if necessary)	
Generator (diesel or gas)	
Grounding/bonding	
Installation	
Land	
Lighting (tower and perimeter)	
Maintenance (first year)	
Permits/filing fees	
Radio system	
Security equipment (cameras, etc.)	
Taxes	
Telephone/data/LAN/video modifications	
Tower/mounting hardware	
Warranty (period)	
Waveguide/coaxial installation	
Total	

line costs from local and long-distance suppliers. From the 5-year costs, a comparison may be made of payout and savings figures (if data can be obtained). However, because of the limited frequency spectrum and the overcongestion in certain areas, there is no guarantee that a path and license will be obtained. Further, some local ordi-

TABLE 2.6 Ongoing Expenses for a Microwave System

Item	Year 1	Year 2	Year 3	Year 4	Year 5	Total
Electricity						
Fees to local municipality						
Floor space						
Insurance						
Maintenance (radio)						
Multiplexer maintenance						
Permit renewal						
Personnel expenses						
Taxes						
Tower fees						
Tower maintenance						
Warranty						
Total						

nances may not allow the installation of a system. Typically, if a T1 (or 24 voice frequency [VF] channels) is required, a fairly quick payback can be accomplished—on the order of $2\frac{1}{2}$ to 3 years. For high-speed digital data transmission needs replacing digital dataphone service (DDS) leased lines, payback periods of less than 2 years can be achieved.

Analog versus digital transmission

Microwave systems are broadly classified as analog or digital, depending on the modulation technique used. Inherently, all microwave transmissions are on an analog carrier system. The input, as either an analog or digital modulation technique, is what makes the difference. Although most older installed systems are analog in nature, the move toward digital has been a major thrust in supplying the higher transmission capacities.

Analog microwave. Older analog microwave systems use either amplitude modulation (AM) or frequency modulation (FM), with FM comprising the bulk of the radio systems installed. In the 30 MHz bandwidth of the 6 GHz common carrier band, 2400 voice channels is the hypothetical capacity of FM technology. Other channel capacities use an amplitude modulation, single sideband (SSB) transmission. With the same 30 MHz, an AM SSB system can carry 6000 channels. Despite the popularity of FM transmission, it is prudent to investigate both amplitude and frequency modulation systems. Each has its own means of modulating (implying) the information onto a radio-based carrier signal.

Amplitude modulation. To understand amplitude modulation, begin by picturing two systems in constant communication with each other via an unmodulated (unchanged) carrier signal. As information is implied on to this unmodulated signal, the amplitude of the carrier signal will change to equal the information being transmitted from site to site, yielding a new signal. The total output of the new signal will have a combination of carrier signals: the original frequency and the differences (sidebands) between the original and new signals. The modulation will produce both high sidebands and low sidebands.

Amplitude modulation can produce several portions of bandwidth. The typical use has been for voice communications, in which 4 KHz of bandwidth is delivered. In Fig. 2.8 this bandwidth is implied (or modulated) onto the carrier signal. The modulation uses a single bit per hertz ratio, or an effective 1:1. This means that for 30 MHz of bandwidth up to 2400 voice channels can be carried, yielding 12.5 KHz of capacity per channel. Clearly, the yield on this is quite low. The frequency is kept constant, whereas the amplitude varies.

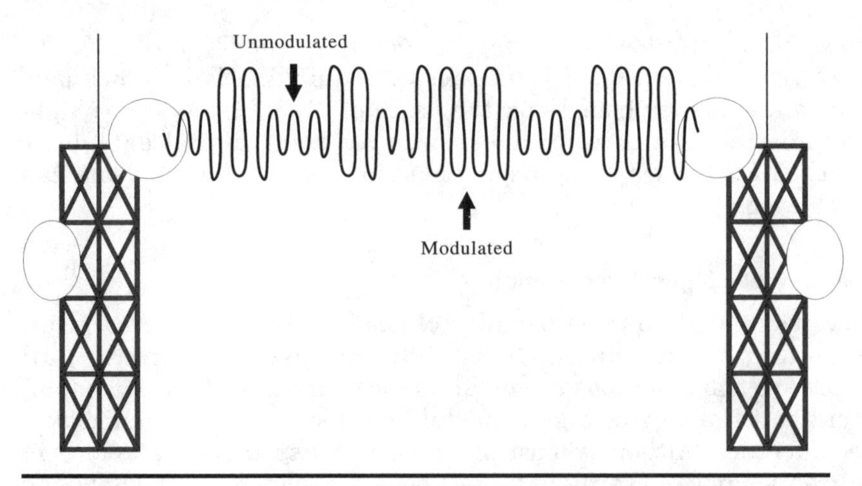

Figure 2.8 The carrier signal is combined with the modulating signal to produce an increase in amplitude, representing the actual data.

Frequency modulation. In frequency modulation, the signal is implied directly onto the carrier using frequency rather than amplitude changes. The unmodulated carrier signal is combined with the modulating signal to produce a modulated carrier signal, comprised of the sum of the frequencies. Figure 2.9 shows the point where the combined frequencies speed up the carrier tone. The amplitude is kept constant and the frequency varies. This is a more acceptable and, therefore, preferred form of transmission.

A block diagram of a complete analog frequency modulation

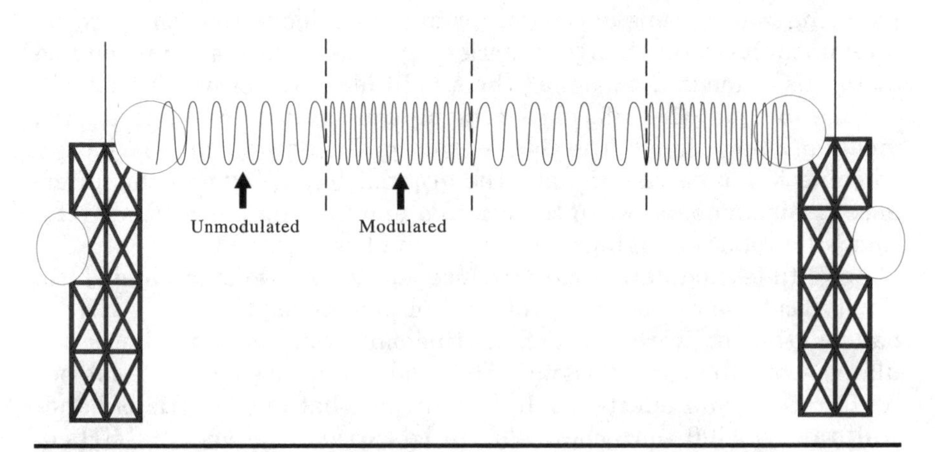

Figure 2.9 In frequency modulation, the unmodulated carrier is combined with the modulating signal to produce a higher frequency rate.

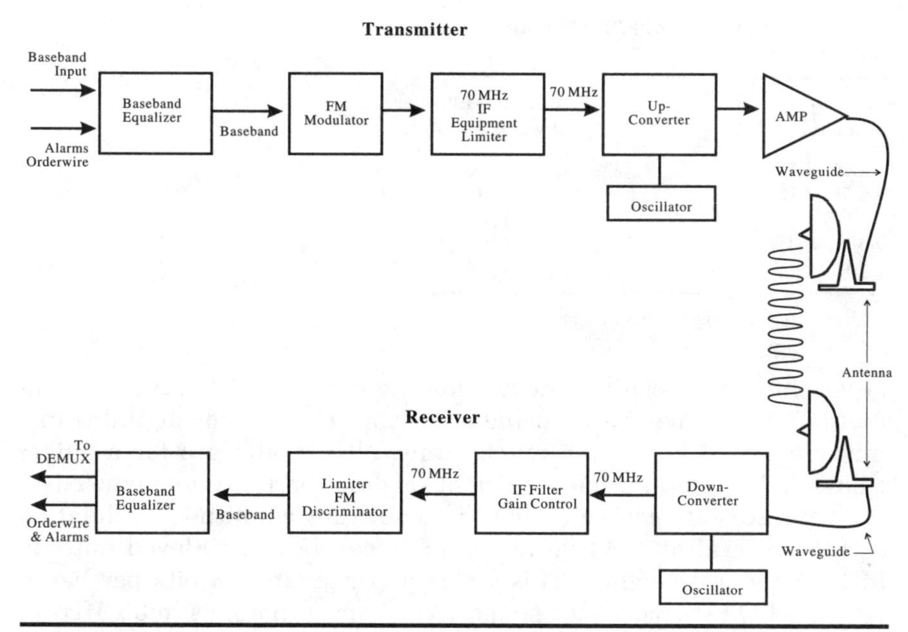

Figure 2.10 A typical microwave analog system.

microwave system is presented in Fig. 2.10. The combined baseband—which includes alarms, order noise, and the frequency division multiplex input—modulates the wave. The modulated signal is converted to an intermediate frequency (IF) of 70 MHz, which is then upconverted to the output frequency being used. The receiver takes the modulated signal and down-converts it to the intermediate frequency of 70 MHz. From the down converter the wave is passed through a filter and amplifier (gain control) to make up for any delay distortion on the filters. The output is then fed to a discriminator for the demodulation process. The original baseband signal is passed into the baseband splitter, which directs the information to the demultiplexer and sends alarms to the alarms and orderwire section of the receiver. Systems using analog transmission can typically carry 24 to 2700 voice channels. On the frequency modulation side, this uses approximately 108 KHz at the low end and 12.3 MHz on the high end.

Digital microwave. Digital microwave has been available since the mid-1970s with direct modulation of the air carrier wave. Therefore, most new line-of-sight systems used for duplex communications are digital. In analog radio systems, one or more of the analog properties (amplitude, frequency, and phase) are quantized by a modulating signal. Since digital implies a discrete set of fixed values, a digital radio system can be represented by a discrete value of amplitude, frequen-

TABLE 2.7 Summary of FCC Minimum Requirements

Frequency range (GHz)*	Minimum number of voice channels
2.11–2.13	96
2.16–2.18	96
3.7–4.2	1152
5.925–6.425	1152
10.7–11.7	1152

*Based on 20 MHz increments.

cy, or phase as a result of the modulating process. With a typical modulation at 1 bit per hertz, using a 64 Kbps channel for digital transmission would be far too much bandwidth to allocate for a single channel. Therefore, a more efficient modulation technique is needed. With phase shift keying (PSK) or quadrature amplitude modulation (QAM), a total of 1344 voice channels can be multiplexed onto 30 MHz of radio frequency. This yields approximately 3 bits per hertz. Newer 64 QAM techniques support 2014 voice channels in 30 MHz of radio frequency. This yields approximately 135 Mbps of throughput (3 T3s) with an efficiency of 4.5 bits per hertz.

Digital microwave systems allow for the direct interface of digital trunks at the T1 and T3 carrier rates. All this efficiency is necessary because the finite resource (the radio frequency spectrum) is limited and already stressed. In the United States, for example, the FCC attempts to make the best use of the spectrum by requiring systems to use a 20 MHz allocation efficiently. Table 2.7 summarizes Part 21.122 of the FCC rules and regulations. Under these requirements,

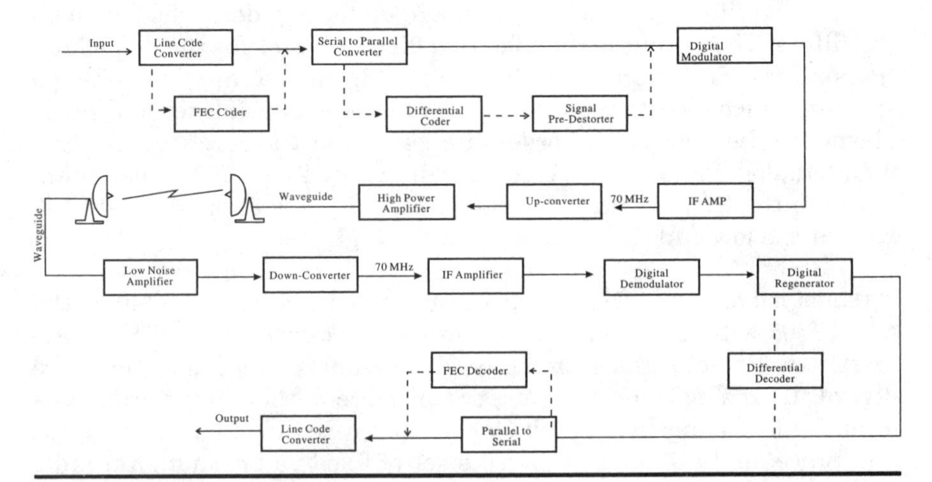

Figure 2.11 A typical digital microwave system.

the bits per hertz may never fall below a 1:1 ratio and the equipment used for voice transmission must meet the minimum number of channels for a specific frequency range.

Figure 2.11 illustrates a digital microwave system. Several of the components shown are optional. At the input side of the radio, a series of baseband functions is accomplished. A line code converter takes a standard pulse code modulation (PCM) line code (such as AMI or B8ZS) and converts it to a non-return-to-zero (NRZ) format. An error detection and correction coder can be used to handle forward errors. However, this function is not always implemented. A serial-to-parallel conversion breaks the serial data stream into two separate inputs—an (I) phase and a quadrature (q) phase for phase modulation techniques. An optional differential coder is used to compensate for phase ambiguity on the receiver end. A predistorter, also optional, compensates for amplification distortion on the signal. The modulator provides the modulation function on the baseband signal. The modulated signal is then converted to an intermediate frequency and is amplified and sent to the up-converter, which converts the baseband to the operating frequency of the radio system. The signal is then passed onto a high-power amplifier (waveguide) and radiated to the antenna for transmission. At the receiver end, the process is reversed so that the information is brought back to its original form.

Phase shift keying is a means of representing the digital transmission to the carrier signal. Figure 2.12 presents a sample of a four-

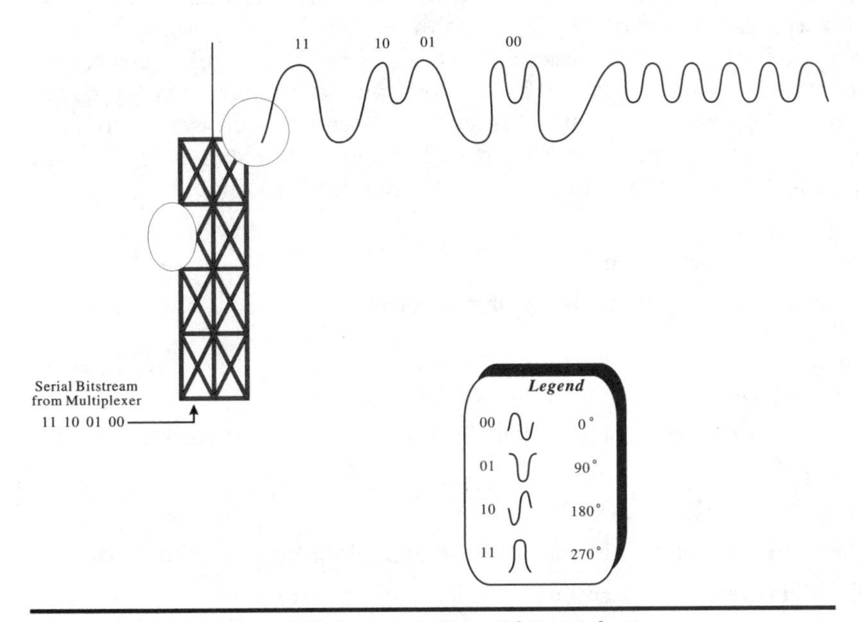

Figure 2.12 A four-phase shift for transmitting 2 bits per hertz.

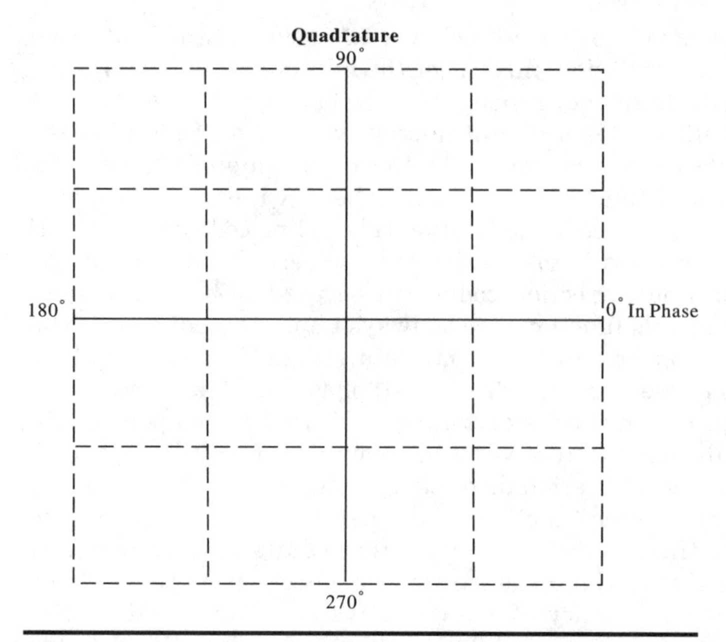

Figure 2.13 A 16 QAM constellation representing the phase and amplitude pattern for transmitting 4 bits per hertz.

phase shift. The shift is represented in its digital form and then delivered to the carrier in its analog form, allowing for the transmission of 2 bits per hertz. Other forms of phase and amplitude shifts can represent more bits per hertz. With 16 QAM (phase plus amplitude modulation) a representation of 4 bits per hertz can be achieved, whereas with 32 QAM a representation of 5 bits per hertz can be achieved. Fig. 2.13 shows 16 QAM in matrix (or constellation) form.

Advantages of microwave

Among the many advantages of a microwave system, the following are distinct:

- Ability to span large bodies of water
- Ability to overcome obstacles (valleys, mountains) in a transmission path
- High bandwidth capacities
- No right-of-way requirement (as with cable) along the entire path
- High return on investment or quick payoff ratio

Disadvantages of microwave

Certain drawbacks and requirements must also be considered:

- FCC licensing
- Path and frequency coordination
- Purchase or rental of real estate for towers
- Power for remote systems
- Equipment security
- Rain attenuation and possible outages
- Disruptions to the line-of-sight path caused by new construction
- Time required to get the service (up to 2 years)

Satellite Communications

In the early days, satellite communications represented a radical change in the processing of long-distance calls. As an alternative to cable for international communication, satellite transmission was very competitive. Further, in rural and underdeveloped areas, it provided quick connectivity solutions for users with a limited number of high frequency (HF) radio channels and allowed large amounts of bandwidth in areas that did not have a cable infrastructure in place. Maritime (ship-to-shore) communications improved through satellite access to the public telephone networks. Finally, satellite broadcasts of high-quality TV signals brought better reception as well as new information access to local communities and entire countries.

Evolution of satellite

As mentioned earlier, the first attempts at using satellite communication occurred in the early 1960s, when scientists began bouncing radio signals off weather balloons and the moon. These efforts met with unfavorable results because of attenuation of the signal as it returned to earth. The earliest satellites were launched in varying orbits that required extensive tracking, with the potential for loss of consistent communications. The design of these systems, which used analog transmission techniques, had to be tightly tuned so that the required quality could be achieved. It wasn't until the mid-1970s, when digital techniques were introduced, and satellites could be positioned in orbits that provided better coverage, eliminated extensive tracking, and improved quality.

Three different orbits are used when deploying satellites:

1. *Inclined orbits.* These elliptical orbits were among the first used and required extensive tracking of the systems. Inclined orbits survive in certain military applications, but they do not satisfy commercial users.
2. *Equatorial orbits.* These elliptical orbits circle the equator. The equatorial orbit has become the lifeblood of commercial users since the introduction of "geosynchronous" orbiting (see below).
3. *Polar orbits.* These orbits revolve around the poles and are non-synchronous. The satellite orbits at various heights above the earth's surface in an inclined path.

Figure 2.14 depicts these three orbits as they would appear above the earth's surface. The choice of orbit determines the type of equipment used and the capabilities to be derived from satellite transmission.

The chief problem with elliptical inclined orbits is that there is only a certain period of time when the satellite is visible. The satellite comes into view on a horizontal plane, passes overhead, and then disappears into the opposite horizon. Thus, the receiver or transmitter station loses contact with the craft. Depending on the orbit used, the height of the craft, and several other variables, timing can be critical.

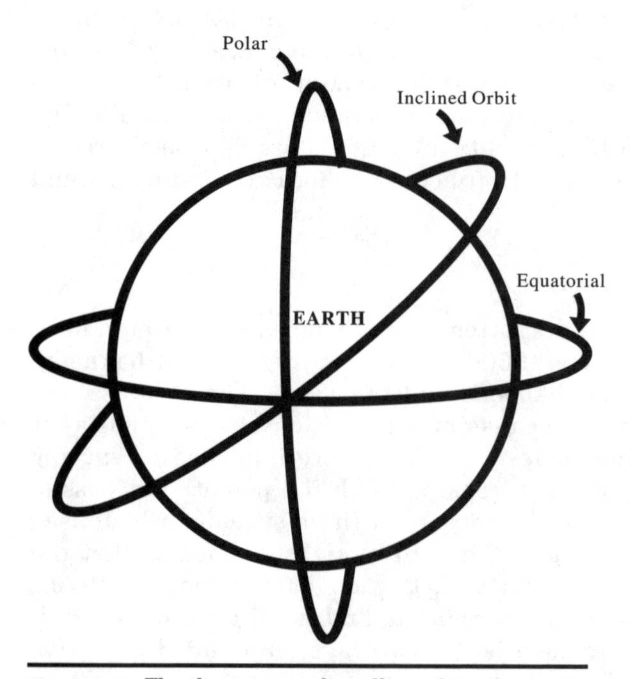

Figure 2.14 The three types of satellite orbits above the earth.

It is not unusual for a tracking station to fail to locate the craft. Predictable locations are used to "lock in" on the satellite as it comes over the horizon. Unfortunately, for whatever reason, the tracking equipment cannot always find it.

Clearly, this process cannot satisfy the demand for constant commercial communications, such as telephony and broadcast TV reception. The solution is a geosynchronous equatorial orbit, with the satellite located 22,300 miles above the earth (at the equator). At this height, the satellite is orbiting around the equator at roughly the same speed that the earth is rotating. Consequently, the craft appears to be stationary and is continually visible to the receiver and transmitter stations. With a geosynchronous orbit, only three satellites are needed to provide coverage of the entire earth.

Figure 2.15 shows the geosynchronous orbit with the appropriate coverage areas. Spaced 120° apart from one another, the satellites provide communications to all parts of the earth's surface except the polar extremes (latitudes more than 81° north or south), with small overlapping areas.

Satellite transmission overcomes the need to use repeaters (approximately every 20 to 30 miles) in long-haul microwave communica-

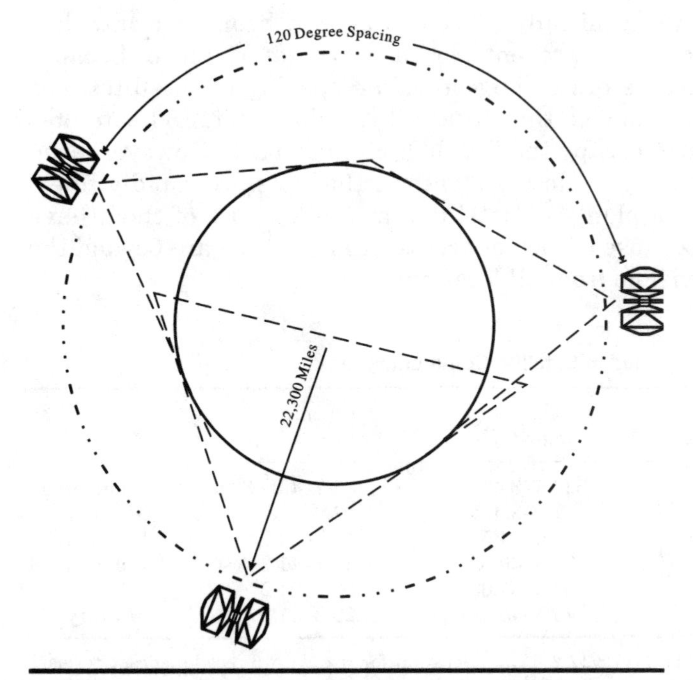

Figure 2.15 At geosynchronous orbits, three satellites are needed to cover the earth's surface.

tions. With the footprint of coverage areas via satellite, the satellite is the repeater. Therefore, on a 3000-mile circuit (or any other portion) only one hop through the satellite is necessary.

Frequency bands for satellite

Like microwave, satellite operates in specific bands (frequency ranges) determined by the Federal Communications Commission (FCC) in the United States and by international regulatory bodies overseas. Satellite bands are typically grouped in pairs such as 6/4 GHz, where the numbers refer to uplink/downlink frequencies. The uplink is the frequency that is transmitted from the radio equipment on the ground, or the earth station to the satellite. The downlink is the frequency used by the satellite to communicate back to earth.

Table 2.8 shows the range of bands assigned for use in satellite communications. Several bands are allocated for transmission, although for commercial use they are limited and many are reserved for microwave terrestrial services. These additional bands are shown in Table 2.9. The primary bands used for satellite, are noted in the table.

Satellite locations

When satellites were initially placed in geosynchronous orbit, they were spaced 4° apart to prevent blockage and interference. Because the segments became quite congested, the spacing of satellites was reduced to 2°. This means that in a full circular orbit, 180 satellites spaced 2° apart can occupy the full 360° circular path. However, even with the craft located so closely together, the slots are rapidly filled. Although there are plenty of satellites in orbit, many of them have excess capacity because of lack of access—that is, they are beyond the horizon and not visible from all locations.

TABLE 2.8 Bands Assigned in Satellite Communications

Frequency	Band	Uplink range (GHz)	Downlink range (GHz)	Use
6/4	C	5.925–6.425	3.7–4.2	Commercial
8/7	X	7.9–8.4	7.9–8.4	Military
14/11	Ku	14.0–14.5	11.7–12.2	Commercial
30/20	Ka	27.5–30.5	17.7–21.2	Commercial
30/20	Ka	30.0–31.0	20.2–21.2	Military
44/20	Q	43.5–45.5	20.2–21.2	Military

Note: a 17/12 GHz (17.3–17.8/12.2–12.7) allocation for use with direct broadcast services has been assigned for future commercial applications.

TABLE 2.9 Microwave Bands by Letter Assignment

Band	Subband	Frequency (GHz)
P		.225
		.39
L	p	.39
	c	.465
	l	.51
	y	.725
	t	.78
	s	.90
	x	.95
	k	1.15
	f	1.35
	z	1.45
	z	1.55
C	e	1.55
	f	1.65
	t	1.85
	c	2.0
	q	2.4
	y	2.6
	g	2.7
	s	2.9
	a	3.1
	w	3.4
	h	3.7
	z*	3.9
	d	4.2
X	a	5.2
	q	5.5
	y	5.7
	d	6.2
	b	6.25
	r	6.9
	c	7.0
	l	8.5/9.0
	s	9.6
	x	10.0
	f	10.25
	k	10.9
K	p	10.9
	s	12.25
	e	13.25
	c	14.25
	u†	15.35
	t	17.25
	q	20.5
	r	24.5
	m	26.5
	n	28.5
	l	30.7
	a‡	33.0/36.0

(Continued)

TABLE 2.9 Microwave Bands by Letter Assignment (Continued)

Band	Subband	Frequency (GHz)
Q	a	36.0
	b	38.0
	c	40.0
	d	42.0
	e	44.0/46.0
V	a	46.0
	b	48.0
	c	50.0
	d	52.0
	e	54.0/56.0
W		56.0
		100.0

*The C band includes frequency range 3.9–6.2 GHz for satellite.

†The K_u band includes 15.35–24.56 GHz for satellite.

‡The K_a band includes higher ranges of frequency for satellite above 30 GHz.

It is not unusual to hear projections stating that the use of satellite transmission is passé. In fact, this form of communication has been steadily expanding. Improvements in technology coupled with the need for high-speed, reliable transmission had been a boon to satellite technology.

Coordination

For the average satellite user, frequency coordination and permits or licenses are nonissues. It is the responsibility of the carrier (owner) to coordinate location, licenses, and frequency spectrum. Then the carrier resells the service to prospective users. In the United States, a number of earth stations (uplinks) are strategically placed around the country. These hubs are then shared among users who lease available space. Unlike microwave systems, which can be purchased (owned) and licensed for private use, satellite systems are too expensive for single organizations. The costs of launch, operations, telemetry equipment for hacking and tuning the orbit, and so on can range from $50 million to $100 million. Few if any organizations would endure or fund this expense for strictly private use.

Access methods

Three methods of access are typically used, two of which are predominant:

Frequency division multiple access (FDMA)

Time division multiple access (TDMA)

Code division multiple access (CDMA)

Derivations of the above access methods include single channel per carrier (SCPC) FDMA, which uses preassigned access, and demand assigned multiple access (DAMA). The following broad summaries of these access methods are provided for illustrative purposes.

FDMA. FDMA—the most commonly used access method for a quarter of a century—will likely stay the primary access method for some time in the future. A single earth station using FDMA transmits a carrier to a satellite. The carrier contains a frequency division multiplexing (FDM) configuration in a modulation envelope. This envelope consists of groups (12 channels) and supergroups (60 channels) assigned to it. A select and unique set of frequencies is used. The transmitters multiplex the information for delivery to the satellite and its transponders for frequency translation and rebroadcast to the receiving earth station. Figure 2.16 shows a typical FDMA. Earth sta-

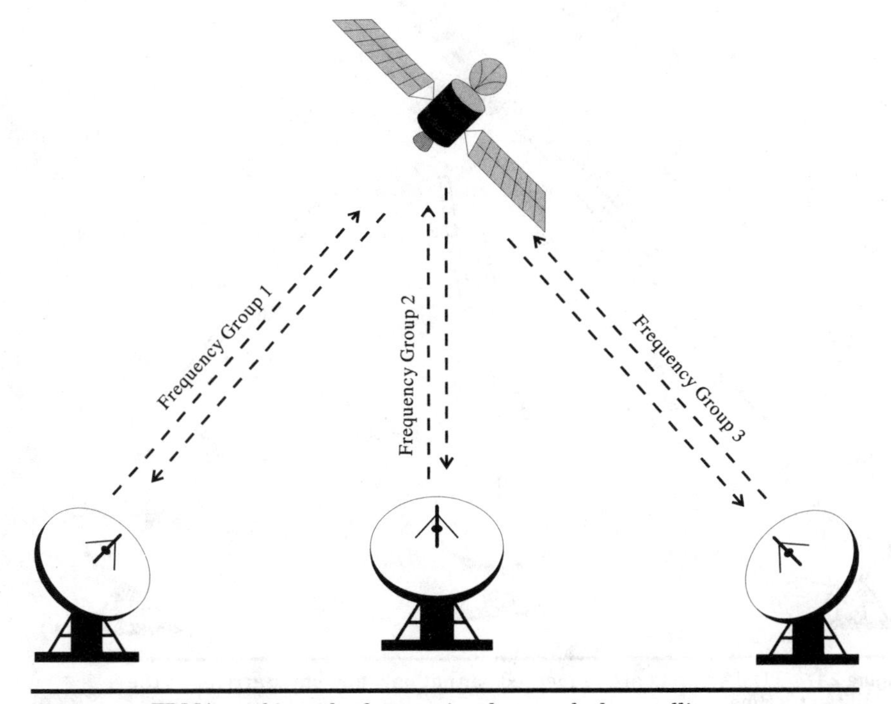

Figure 2.16 FDMA combines the frequencies that reach the satellite as a composite signal.

tions sort out the multiplexed signal by tuning to a particular frequency and processing the information accordingly. FDMA is fairly mature and straightforward to use.

TDMA. TDMA is the second most common form of satellite uplink connection. Unlike FDMA, TDMA uses transponders that receive bursts of data from the transmitters. These bursts arrive sequentially (one at a time) at the satellite. Therefore, no conflict arises in the modulation between the transmitters and receivers. TDMA operates in a time domain and is thus applicable to digital systems, since buffering of the information is required. Each station has a buffer that stores the output. With control coming from a central site, the individual buffers transmit the information in very high bit rates via the uplink and into the satellite. Depending on the bandwidth of the transponder and the modulation scheme used, bit rates of 10–100 Mbps can be transmitted. Figure 2.17 illustrates the bursts of TDMA data. TDMA uses the bandwidth far more efficiently than FDMA. With a standard FDMA service from an uplink,

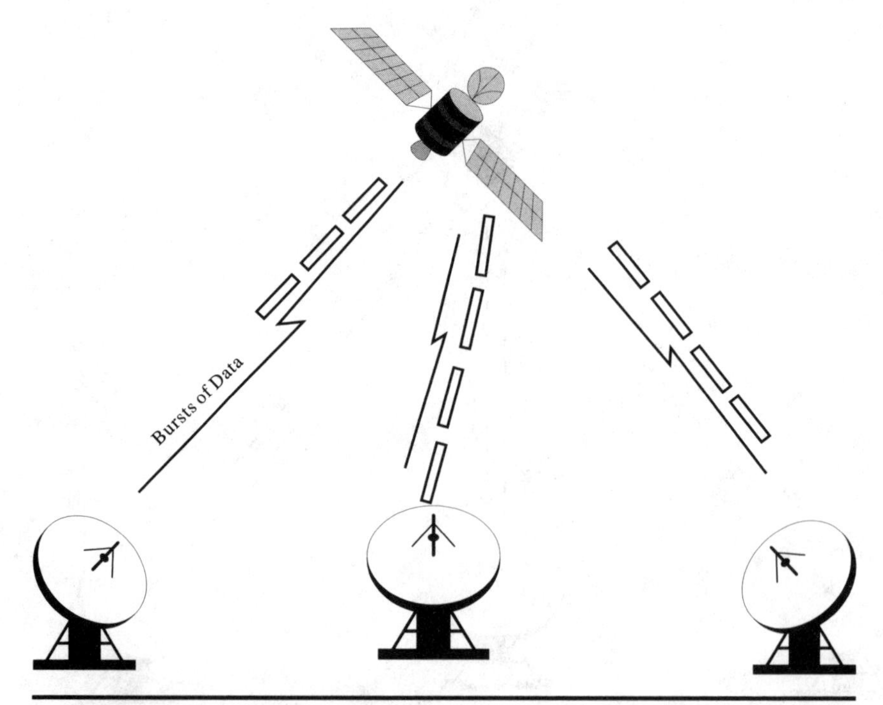

Figure 2.17 TDMA bursts are sequenced so that only one burst arrives at the satellite at a time.

approximately 450 channels can be achieved; with TDMA and 64 Kbps PCM (the digital time division multiplexed signal) voice transmission, 1800 channels can be achieved. The benefits of 64 Kbps PCM on TDMA are evident.

CDMA. CDMA is relatively new and not widely accepted or implemented. It has been used by the military for a long time but has limited deployment in the commercial arena. With CDMA, all stations transmit on the same frequency simultaneously. To prevent chaos, each transmission has its own code, which cannot be decoded unless the receiver has suitable equipment to detect and reproduce the original information. Even if received, interfering information has no effect, since it is not within the limits of a specific code. Each receiving terminal has its own code, rejecting all others. Included in CDMA is a technique known as *frequency hopping,* in which one and then another frequency carries the information. CDMA is expensive because of the complexities of the coded chips and frequency-hopping equipment. Figure 2.18 illustrates the CDMA technique.

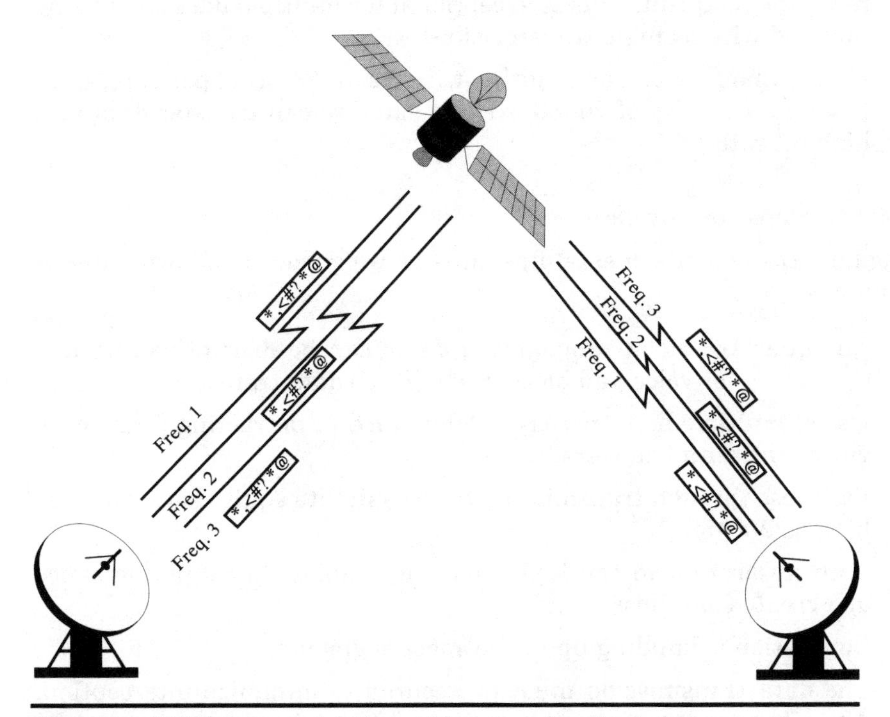

Figure 2.18 CDMA uses a coded chip and frequency-hopping techniques.

Advantages of satellite

Satellite transmission offers the following advantages:

- Distance insensitivity—for long-haul communications the transmission path of 22,300 miles up and back from the satellite makes terrestrial distances a moot point.
- Single-hop transmission—a transmission of several thousand earth miles may be accomplished using a single repeater (the satellite). With microwave, the same transmission would require numerous terrestrial repeaters.
- Improved communications—a satellite brings connectivity to remote areas and to developing countries where a cable infrastructure doesn't exist or would not be feasible. Ship-to-shore communications are also improved.
- Good error performance for data transmissions—since the signal leaves the earth's atmosphere, the bit-error performance should be $\approx 10^{-10}$.
- Broadcast technology—a single transmission can be broadcast to multiple receivers simultaneously so long as they are within the receiving footprint. This is exceptional for multiple data sites being updated with a single transmission.
- Large amounts of bandwidth—with 12 to 24 transponders, each having a capacity of 20–36 MHz, a satellite can transfer data at a high bit rate.

Disadvantages of Satellite

Against the benefits of satellites must be weighed the following disadvantages:

- The quarter-second propagation delay can be disruptive and disconcerting for voice and block mode (BSC) data transfers.
- Use of multiple hops increases delay, further detracting from voice, video, and data transfers.
- Path loss through transmission to the satellite equivalent can be as high as 200 db.
- Rain absorption in the high-frequency range affects performance and creates path loss.
- Congestion is building up in the space segments.
- The data transmission must be secured to minimize interception. Any site tuned to a specific receiver frequency could intercept the information without detection.

Chapter

3

Light-Based Systems

In wireless communications, the primary means of transmitting light is through infrared light in free space. Operating in the frequency spectrum of laser beams in the terahertz (trillions of hertz per second), an invisible infrared light beam is focused from a transmitter to a receiver over a very short distance. Infrared light is subject to even more specific curtailments than the radio technologies, such as:

- Transmission distances of less than 2 miles
- Line-of-sight limitations
- Bandwidth restricted to today's maximum of 16 Mbps throughput
- Proneness to environmental disturbances
- Disruption from fog, dust, heavy rain, and objects in its path

No single wireless technology can be 100 percent effective; each has its own limitations and strengths. Despite the technical bandwidth limitations, there are certain cost advantages and ease-of-use considerations that keep light-based systems high on the scale of acceptability in the minds of many users for a variety of applications. Only a few developments in technology have sparked the interest and creativity of developers and users alike. One such development has been laser transmission. The availability of low-powered solid-state laser diodes has opened up a myriad of possibilities for producing cost-effective short-haul communications links through the air to meet such transport needs as voice, data, video, and now LAN traffic at native speeds.

Dynamics of Laser Transmission

An atmospheric laser transmission system helps overcome the stringent limitations and obstacles associated with other types of short-haul communications. Transmission through the air, without the use of an infrastructure, reduces the need for a right-of-way, as is necessary with a cable system. Further, infrared laser systems fall under the domain of nonlicensed technology, thereby removing many of the licensing requirements and associated delays and costs. For under the 2-mile range, this technique holds a wide range of exciting possibilities. Several criteria must be considered before adopting such a system:

The basic geometry of the system

Atmospheric conditions that will affect the transmission

Site selection for installation of the link

The system geometry

As already mentioned, an infrared laser system is generally employed in a point-to-point, line-of-sight application. Some multipoint and repeatered systems have been developed, but these are few and far between. Three elements make up the basic system:

- The transmitter station, where the signal gets created or modulated from its original form into a light beam
- The medium used—in this case, air
- The receiver station, where the information is put back in its original form

The basic components of the laser system are shown in a gun format in Fig. 3.1. The data are input into the transmitter and modulated onto a light beam. The raw laser is a very narrow beam of light with very little divergence. Its properties are referred to as *monochromatic.* Theoretically, the beam is usable to the receiver in its raw form, but in reality it is extremely difficult to aim and maintain the beam. Therefore, the raw light must pass through a lens to produce a divergent beam. Although this helps solve the initial limitation of the beam itself, it also creates some of the distance limitations of the transmitted signal, since the lens reduces the distance from the transmitter to the receiver by diverging the beam. Because the safety requirements limit the total output power of the beam, the transmitter generally uses a constant diameter of the signal, called a *footprint,* regardless of the distance involved.

The received light is focused by a collector (collecting lens) at the receiver, which then sends the received signal to an optical detector. The receiving lens has a narrow focus—a receiver angle of acceptance (RAA) of approximately 3–5 mR—to provide a degree of selectivity.

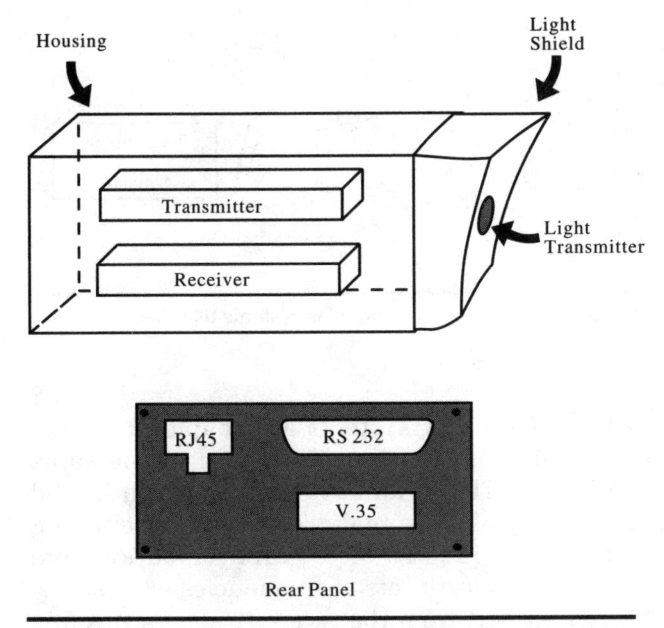

Figure 3.1 The basic components of an infrared system.

The transmit beam divergence angle (TBA) is always greater than the receiver angle of acceptance. The receiver needs a minimum level of incident light in order to demodulate the beam back to the original input signal. There are tradeoffs in the size, complexity, and cost of the receiver lens with its ability to collect the light, thereby limiting the range. The receiver angle of acceptance and the transmitter beam divergence angle are shown in Fig. 3.2.

The above discussion applies to a simplex (one-way point-to-point) system. However, most applications need the transmission and reception of information. Therefore, a duplex system is required. To

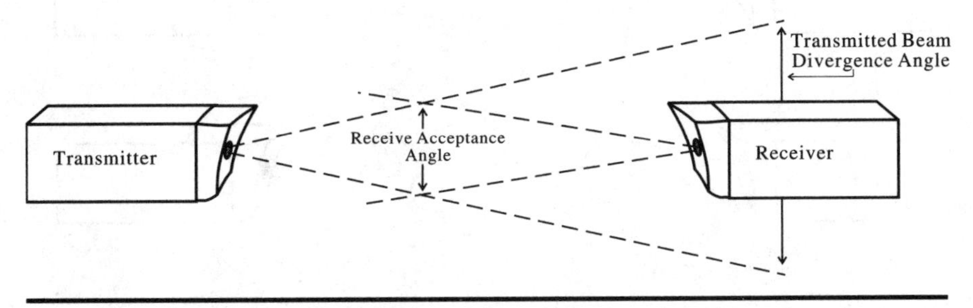

Figure 3.2 The transmitted beam divergence angle compared to the receiver acceptance angle.

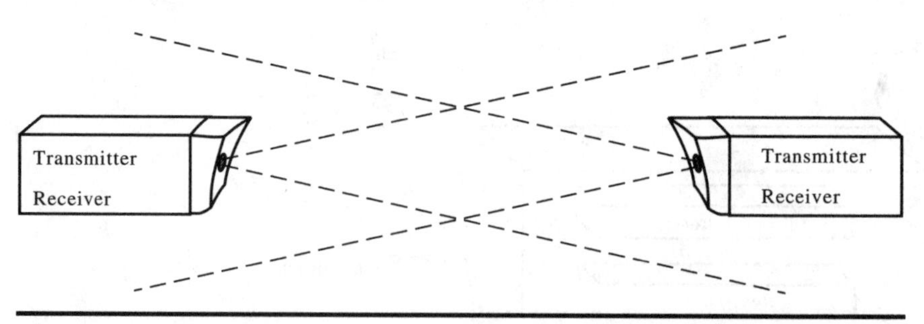

Figure 3.3 A duplex system is created by linking two simplex systems together.

create a duplex system, two simplex systems are coupled together (see Fig. 3.3). Both systems will operate with lasers at the same wavelength (typically 830 nm). This is an advantage of a light-based system, since the directionality of the signal does not migrate from a transmitter to a receiver, as is true of a radio-based system. Remember that in radio two separate sets of frequencies are used to keep the transmit frequency apart from the receive frequency. A further feature of focused-light transmission is that multiple systems can be mounted close together without causing interference so long as the angle that the systems make with one another is greater than the original angle of the receiver. In the multiple systems shown in Fig. 3.4, two light beams are literally transmitted across the same path, yet are kept separate because of their different angles of acceptance.

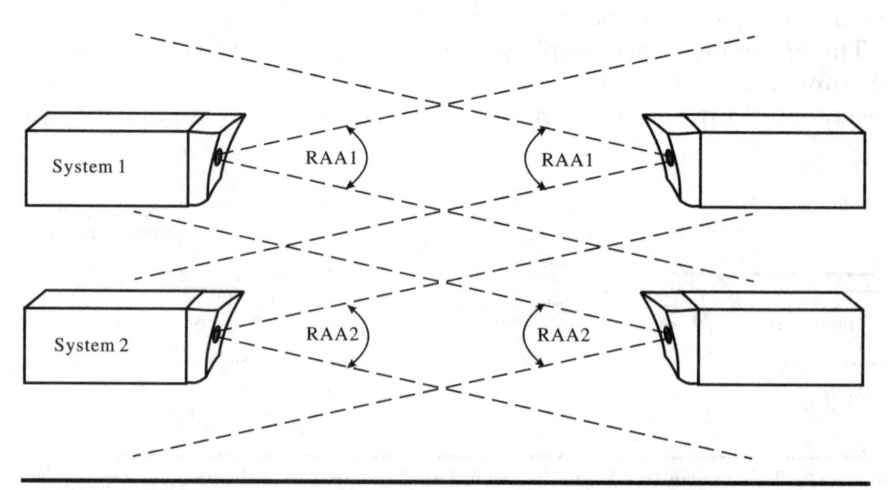

Figure 3.4 Multiple systems can be installed close together without conflict.

Atmospheric conditions

The influence of the atmosphere on a light transmission will affect the overall performance of the system. The severity of the atmospheric conditions and the length of their effects will degrade the distance and performance accordingly. The positioning of the equipment will also be somewhat affected. As noted in the discussion of radio-based systems in Chap. 2, all electromagnetic radiation used in a communications system is affected by the atmosphere. However, each type of transmission system is affected differently, whether it is low-high frequency radio, microwave radio, or laser. The wavelengths of each of these systems determine the degree to which the atmosphere affects the actual transmission.

The three most significant conditions that affect laser transmission are absorption, scattering, and shimmer. All three can reduce the amount of energy at the receiver, thereby compromising the overall reliability of the system and the bit-error levels of the received information.

Absorption. Absorption is caused primarily by the water vapor (H_2O) and carbon dioxide (CO_2) in the air along the transmission path. Their presence is a function of both humidity and altitude. Gases in the atmosphere have many resonant bands, called *transmission windows,* which allow specific frequencies of light to pass through. These windows occur at various wavelengths; the one we are most familiar with is that of visible light. Because the near-infrared wavelength of light (830 nm) used in laser transmission occurs at one of these windows, absorption is not generally a big concern in an infrared laser transmission system.

Scattering. Scattering has a greater effect than absorption. The atmospheric scattering of light is a function of its wavelength and the number and size of scattering elements in the air. The optical visibility along the path is directly related to the number and size of these particles. *Optical visibility* is a meteorological term describing the distance at which the human eye can distinguish a 1-meter-square black target against a white background. The definition implies that visibility is a matter of contrast. The most common scattering elements in the air that affect laser beam transmission are fog and smog, rain, and snow.

Fog and smog. Fog appears when the relative humidity of air is brought to an appropriate saturation level. Some of the nuclei then grow through condensation into water droplets. Attenuation by fog is directly attributable to water droplets less than a few microns in radius. Fog produces a scattering effect in all directions because of the size of the water droplets and the wavelength of the laser beam. The result of the scattering is that a smaller percentage of the transmitted light reaches the receiver. Visualize a moving car caught in dense fog. The headlights on the car are rendered somewhat useless. Turning on the high beams simply scatters the visible light from the headlights, making it almost impossible to see. The light is being reflected back at the vehicle as well as being scattered in all directions. This is exactly how fog affects the infrared light beam being transmitted.

Smog produces similar effects to fog but to a lesser degree, because the radius of the particulate matter is larger. Consequently, fog is the limiting factor in laser transmission performance. The visibility-to-attenuation ratio required for error-free performance varies with the length of the path and the amount of particulate matter in the atmosphere.

Rain. Although the liquid content of a heavy shower is 10 times that of a dense fog, the radius of a raindrop is about 1000 times that of a fog droplet. This is the primary reason that attenuation via rain is 100 times less than that of fog. Think about the difference for a moment. When fog is created, the saturation level is very high in the air, but the droplets are relatively fine. In rain, although more liquid is falling, the drops of liquid are coarser. Because there is more space between the drops of rain than between the droplets of fog, light can get through rain much easier. Let's return to our example of a moving vehicle. In a heavy rainfall, the light still penetrates the raindrops and gives the driver greater visibility over a greater distance. When the high beams are turned on, the distance is better than with the low beams. A certain amount of the light is still being reflected back at the vehicle and scattered in all directions, but more of the light is penetrating through the rain. Similar effects occur with the transmission of a laser beam through the rain. It would take an extremely heavy rain (greater than 5 inches per hour) to significantly reduce the laser transmission.

Snow. The effects of snow on a laser transmission fall somewhere in between those of fog and rain, depending on the degree of water particles in the snow. A very wet snow is close to rain, whereas an extremely dry snow is similar to fog. The key factor, again, is the radius of the particulate matter of the snow. The radius of a wet snow emulates that of a raindrop, whereas the radius of a dry snow is closer to that of a fog droplet. Once again, try to visualize driving through a dense snowstorm. When the snowfall is wet, the experience is like driving through a heavy rainfall. It is not all that pleasurable, but at least the light penetrates the snow and visibility is achieved. However, when a "white-out" occurs, it is much more difficult for the light to penetrate through the snow. The light appears to be reflecting right back at the vehicle, rendering all visibility useless. The same effect occurs on a laser transmission.

Shimmer. Picture driving down a long road on a hot day. As you glance toward the horizon, it appears that water is on the road. As you look closer, you see what appear to be heat waves shimmering off the ground. This is the direct result of a combination of factors, including atmospheric turbulence, air density, light refraction, cloud cover, and wind. The combination of factors will cause a similar disturbance when a laser beam is transmitted through the atmosphere. As the local conditions in the area begin to combine, shimmer will impose a low-frequency variation on the amount of light being transmitted to the receiver. This variance could result in excessive bit-error rates or video distortion on a laser communication system. Simple, proven techniques such as frequency modulation (FM) and automatic gain control (AGC) can minimize the effects of shimmer on the system. Using a path several meters above the heat source will also reduce the negative effects of shimmer.

Site selection

To ensure successful installation and use of a laser transmission system, several site factors must be taken into consideration. These include, but are not limited to, clear path, mounting conditions, and mounting structures.

Clear path. When a transmission system is set up between two points, a clear line of sight is essential. This means that nothing can lie in the way of the light passing along the path. The two ends must physically "see each other" or the system will not work. It is as simple as that. The laser beams should not be directed through or near wires (electric, telephone, cable TV, etc.) or through tree limbs. Depending on the location selected, provisions should also be made to prevent the nesting of birds in the path.

In Fig. 3.5 a clear line of sight is emphasized as the optimal means of setting up the shot between two sites. In Fig. 3.6 care must be taken to get over cables and wires along the path, and in Fig. 3.7 allowances must be made for the growth of tree branches over time. As shown in Fig. 3.8, the system should be mounted at least 10 feet above pedestrian or vehicular traffic to prevent direct eye contact with the laser and disruption of the signal. Unusual circumstances

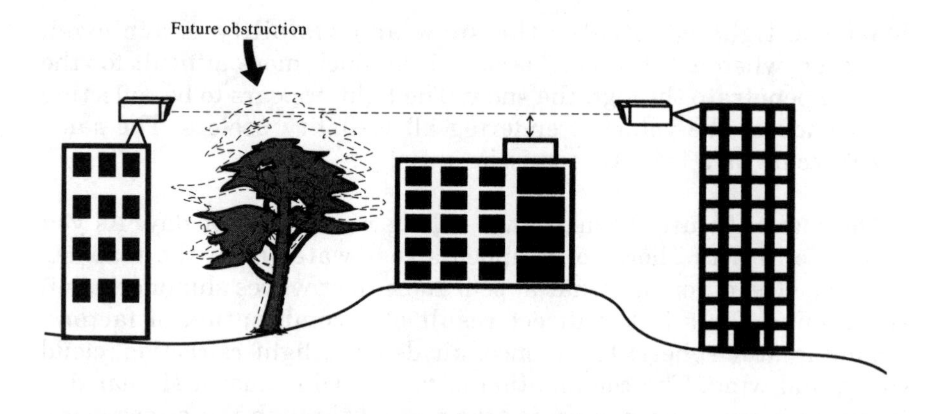

Figure 3.5 A clear path must be ensured for the transmission.

such as exhaust fumes from aircraft and automobile traffic must also be accommodated.

Systems can be mounted either indoors or outdoors. In Fig. 3.9, the system is mounted indoors with the beam passing through glass. However, care must be taken to keep the laser perpendicular to the glass to prevent any angle of reflection from the glass itself. Further, the glass cannot have any reflective coating or tinting. A reflective coating may impair the beam's ability to pass through it in either direction. Although not a major consideration, window-cleaning personnel should be kept away from the area. Looking directly into the laser will cause severe damage to their eyes. Also, they will block the signal for the duration of their cleaning efforts, and they could leave window streaks that reflect or refract the light away from the

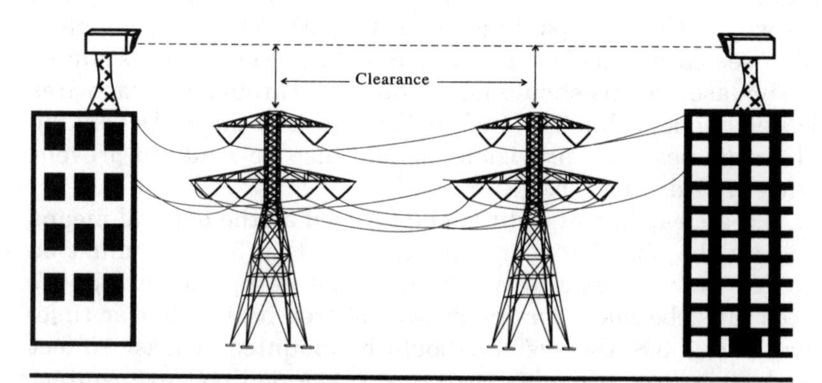

Figure 3.6 Towers may be necessary to keep the path above electric or telephone wires.

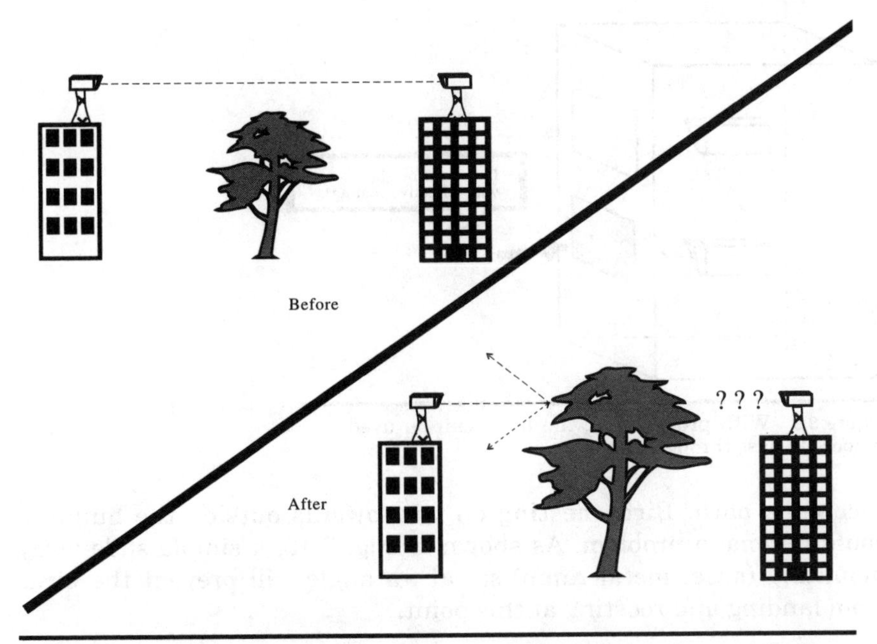

Figure 3.7 Allowances must be made to get around tree limbs, so that growth of limbs will not block the signal.

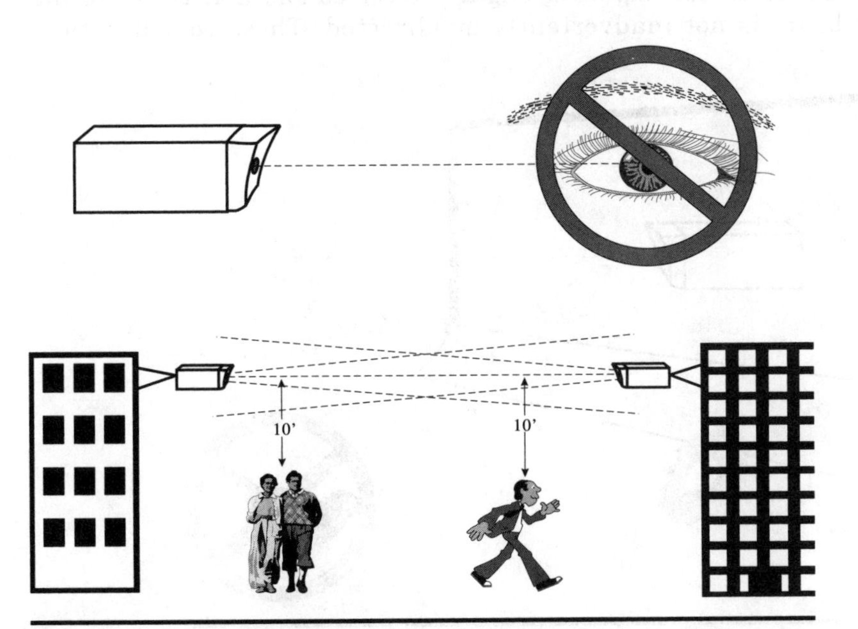

Figure 3.8 The system should be a minimum of 10 feet above pedestrian traffic to prevent direct eye contact.

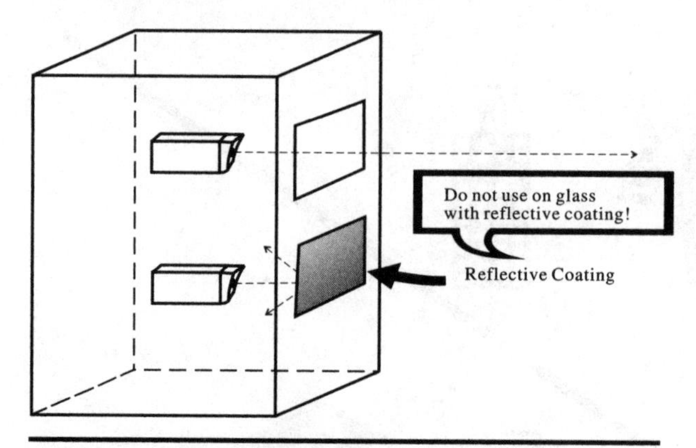

Figure 3.9 With proper care, the laser can be used indoors to pass through glass.

receiver's path. Birds nesting on windowsills outside the building could become a problem. As shown in Fig. 3.10, a simple slide-away mounting (sheet-metal ramp) set at an angle will prevent the birds from landing and roosting at this point.

Mounting conditions. A few precautions must be considered when determining where a laser transmission should be mounted to ensure that the maximum signal strength is received and that the transmitted beam is not inadvertently misdirected. These considerations

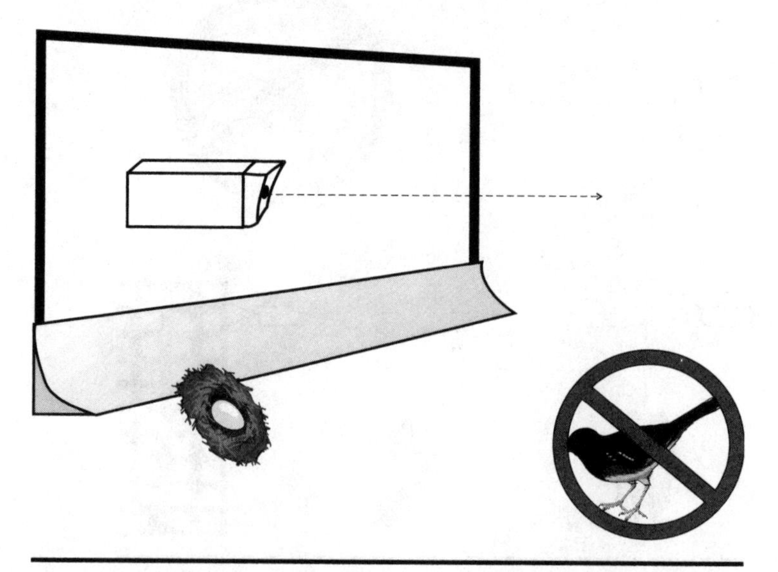

Figure 3.10 A slide-away mounting can prevent birds from nesting in front of a system.

include the avoidance of heat sources, water obstructions, east-west orientations, and vibration.

Heat sources. Heat scintillation (or shimmer) occurs as a result of differences in the index of refraction of air with temperature. The air itself may act as a lens that incorrectly steers the laser beam. Asphalt roofs, parking lots, and similar surfaces may reflect a considerable amount of heat that could cause localized changes in the air temperature. This in turn influences the beam of light along the path. The units should be mounted at least 20 feet above such reflective surfaces to prevent misdirection of the beam. In addition, the beam should not be directed over exhaust vents, cooling towers, chimneys, smokestacks, or large air-conditioning units. Selecting a leading edge of a roof will help eliminate these types of obstructions and also prevent snow buildup in areas where snow can be a problem. Figure 3.11 summarizes these considerations.

Water obstruction. Water droplets on glass can act as a lens that redirects the light. To prevent this from occurring, mount the system behind a glass window and hang an awning above. If building codes or other situations prevent use of an awning, mount the unit as high as possible on the glass. This will prevent the buildup of droplets on this section of the glass, since the water will run down quickly.

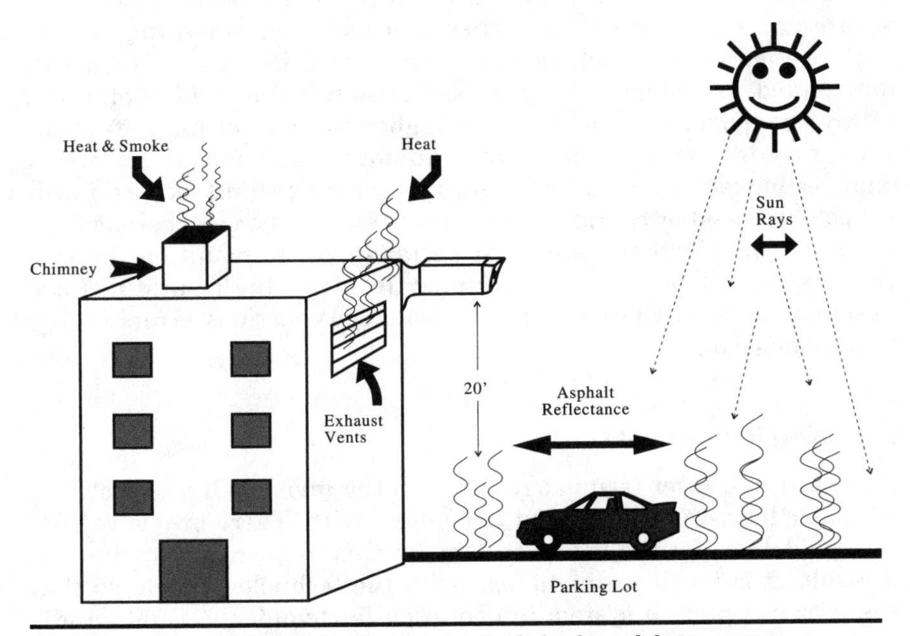

Figure 3.11 Care should be taken not to pass the light through heat sources.

East–west orientations. The angle of acceptance in a laser system is designed to be quite narrow to minimize the acceptance of another infrared light. Any extraneous infrared light passing directly into this angle of acceptance will obviously be accepted, thus causing errors. Of particular danger is an east-west shot that will have direct sunlight on the same axis of the system. The sun contains infrared light in its total light source, which cannot be filtered out. Therefore, a system will be "blinded" for periods when the sun's axis is on the same axis as the light beam from the system. Since an outage of the communications system will result, this orientation should be avoided whenever possible.

Vibration. Since both the transmitter and the receiver deal with very narrow beams, the smallest amount of movement of these units can severely limit transmission or reception. Alternatively, the system could lose its angle totally, resulting in an outage. Keeping the systems away from vibrations of machinery, air-handling units, and other equipment is essential.

Mounting structures. Transmitter and receiver systems are typically designed to project a beam footprint 6 to 8 feet in diameter at the receiver, and have some latitude for minor beam movement. However, movement of the system should be kept to a bare minimum to ensure proper performance. The units should be mounted on a secure surface, such as the corner of a building. Mounting on any surface that sways—such as a tree, fence, or utility pole—is not recommended. Obviously, the greater the movement (which may not always be evident to the eye), the higher the risk of data errors or total system disruptions. A less obvious situation arises when changes in weather conditions cause a wooden mount (such as a utility pole) to swell or contract. This movement can be treacherous for a laser system. Further, care must be taken when mounting the system inside tall buildings. The upper levels of high buildings are designed to sway. Once again, this could cause serious errors or loss of transmission.

Licensing Requirements

Since infrared laser technology relies on the invisible light spectrum, no major licensing issues need to be dealt with. There are no regulatory guidelines for the use of systems in this frequency spectrum. As a result, it is much easier to deal with the technology itself, so that the user can place a system in operation fairly quickly. Unlike radio technology, in which the bulk of the frequency spectrum requires licenses and path clearances (with limited exceptions) from regulatory bodies around the world, the use of light will theoretically cause no interference between systems. Thus, any use of this spectrum is allowed with little to no coordination effort.

As noted earlier, two light-based systems can be mounted side by side, or can even cross each other's path. So long as the receiver acceptance angle does not coexist on the exact same axis, there should be no problems. This is an attractive feature of laser technology. Keep in mind, however, that distances are limited. Also, manufacturers of light-based systems must abide by certain constraints regarding radiation and must issue the necessary warnings that direct contact with the output beam must be avoided. After all, this is a laser and precautions must be taken to avoid risk of direct exposure. The manufacturers go through a certification process with the National Center for Devices and Radiological Health, a division of the Food and Drug Administration, to prove that their units are safe. From the user's perspective, checking that a unit bears the label proving it is certified safe is an appropriate step.

Bandwidth Capacities

The bandwidth of light-based systems can cover a myriad of capacities depending on the configuration purchased from a vendor. The specific application must be considered in terms of the bandwidth necessary as well. At the high end of the bandwidth, systems today can deliver as much as 16 Mbps throughput for LAN-to-LAN connections on a 16 Mbps token passing ring (TPR) network. Thus one could say that the bandwidth is 16 Mbps. However, other systems are designed to meet the needs of the user interface. Table 3.1 summarizes the various features. The needs are met by different systems at the capacities appropriate to a given application. In several of these systems, the pricing is dependent upon actual needs.

TABLE 3.1 Light-Based Systems and Their Applications

Bandwidth	Application	System name
10 Mbps	Ethernet data link	LOO-28
4 and 16 Mbps	Token passing ring	LOO-59F
1.544 Mbps	T1 capacities data/video	LOO-36F
4 X 1.544 Mbps	Quad T1 voice/data	LOO-56F
RS-232–19.2 Kbps	Data transceivers	LOO-12
RS-422A–2.048 Mbps	Data transceivers	LOO-13
Audio 20–20 KHz	Audio transceivers	LOO-15

SOURCE: Laser Communications Inc., Lancaster, PA.

Applications

As noted in Table 3.1, infrared laser technology has several applications. These are addressed below as they pertain to filling a short-haul need for connectivity from site to site. In the scenario presented, a typical customer with two sites (buildings) located approximately 1000 meters apart needed to provide various forms of connectivity between the sites. After reviewing the situation, the customer considered these choices:

A direct-burial cable (copper, coaxial, or fiber) system

A microwave radio system

An infrared laser system

Although more consideration and information would be needed to arrive at a sound conclusion, the following scenario serves as a useful comparison.

The Cable Decision Upon looking at the cable alternative, the customer was faced with extraordinary costs to dig a trench, lay a conduit, and run a cable connection to the two buildings. In addition, the necessary right-of-way could be extremely difficult to obtain. The customer decided that a cable-based solution was not appropriate in this situation. Another alternative was to lease line facilities from a local exchange carrier (LEC) or a competitive access provider (CAP). However, the long-term leased-line costs were not attractive to the customer.

The Radio-Based Decision Looking at the area and the potential ordinances and variances involved, the customer decided to reserve the microwave system as a second choice only if another alternative did not work. Because certain forms of licensing would be required and a clear path would have to be established in a very heavily installed radio area, the costs and time constraints were greater than the customer wished to endure. Further, there was no guarantee that a license would be granted, even if the customer invested the required time and effort. Hence, the microwave choice was all but ruled out.

The Infrared Laser Decision Since the cable and microwave choices were not especially attractive, the customer turned to the infrared laser alternative. No licensing requirements were necessary, so the system could be put into action as soon as the acquisition was made. Since no right-of-way constraints were involved, the process could be handled quickly. An added benefit was that there would be no recurring monthly charges for use of the facilities, as would be the case with leased lines. Lastly, implementation required no major power equipment, towers, cable entrances, or other construction. The purchase price of the system was thus very attractive: a one-time cost of less than $20,000. The customer also took into consideration that if one of the sites ever had to be abandoned, the system could be packed up and moved along with the rest of the office equipment. Assuming that a clear line of sight could be obtained at the new location, and that the distance did not exceed the specifications, the system

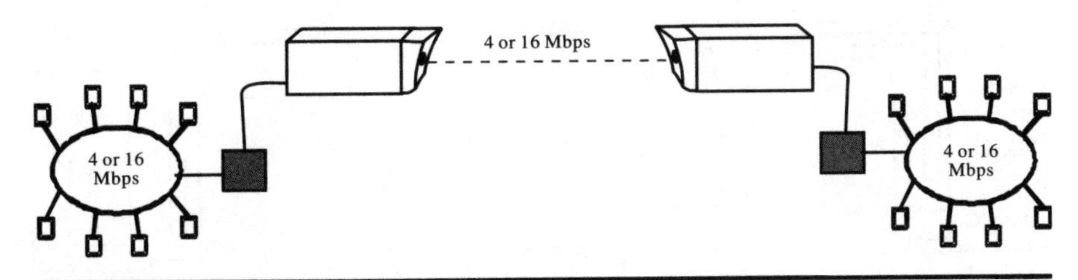

Figure 3.12 Linking two token-ring LANs together at a native speed of 4 Mbps or 16 Mbps provides transparent connectivity.

could be put back into action in a matter of hours. This would not have been the case with the other options. The infrared laser system was the clear choice.

But what of the application? Regardless of the application, the same logic would have governed the customer's decision. This particular scenario might have encompassed the following applications.

Token Ring to Token Ring Infrared laser technology can link two token-ring LANs together at native speed, as shown in Fig. 3.12. Either a 4 Mbps or a 16 Mbps capacity can be applied. Native speed means that the LANs are connected at the same speed at which they operate. This allows for transparent connectivity at the full-rated speed of the network itself—for approximately $35,000 at 16 Mbps or $26,000 at 4 Mpbs. If the transport were slowed down to some lower-level speed, through a bridging arrangement, the total throughput would be limited to a speed of 1.544 Mbps or less.

Ethernet to Ethernet Infrared laser technology can link two Ethernet LANs together at a native speed of 10 Mbps, as shown in Fig. 3.13. A different system is used here, with the native throughput operating at the full-rated 10 Mbps. The cost is only about $15,000. Again, a slower-speed bridging arrangement would be limited to 1.544 Mbps or less.

In both of these cases, an alternative could be to lease T1/E1 speed lines at slower connections than the LANs operate. Network congestion might result, since the throughput must be buffered to accommodate the line speed. The

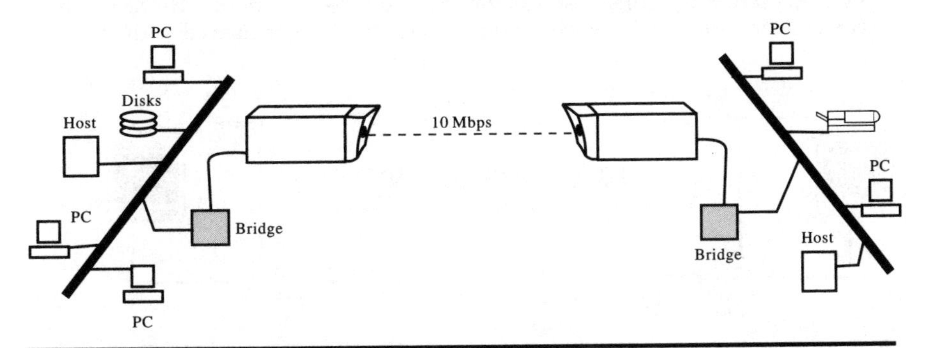

Figure 3.13 Infrared can link two Ethernet LANs together at a native speed of 10 Mbps.

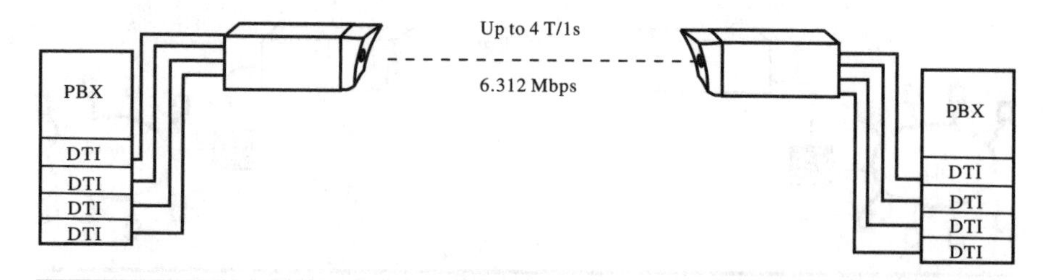

Figure 3.14 Linking PBXs together with up to four T1s for TIE lines. DTI cards are used in the PBX to yield up to a 96 PCM circuit at 64 Kbps.

leased line of up to 1 mile from a local operating company (telco) could cost as much as $700–$1000 per month. The payback on this arrangement would obviously be dramatic, gaining the additional speed along with a return on investment in 2 or 3 years, depending on the system selected. Further, if the native speed were leased from the local telco, the cost for the service could range from $1800 to $2400 (depending on the speed selected), thereby producing a return on investment in 12–16 months for close-in services. By way of contrast, complete installation of a microwave system could cost $80,000 to $100,000, or three to four times the cost of infrared installation. As can be seen, for this environment an infrared laser system has some definite advantages.

PBX to PBX Figure 3.14 shows the connection of PBX to PBX using a digital transmission system for TIE lines between systems with a quad T1 capability for voice connectivity. A digital trunk interface connection is used at each end of the system. In this particular situation, 56–64 Kbps of voice connectivity are provided with 24 channels on each of the four T1s, resulting in 96 digital voice connections. This is a prime arrangement for connecting two PBXs in close proximity to a single network or for providing other voice-networked services (voice messaging, off-premises extension, etc.). The infrared laser system would price out at approximately $23,000, yielding a net payback in less than 1 year, given four T1s at $1000 per month rental from the local telco.

T1 to T2 Figure 3.15 illustrates the provision of a T1/T2 service between two sites to provide a 6 Mbps channel capacity or four 1.544 Mbps channels for video-conferencing capabilities at compressed video standards. However, these compression rates will deliver superior-quality video compared with the even

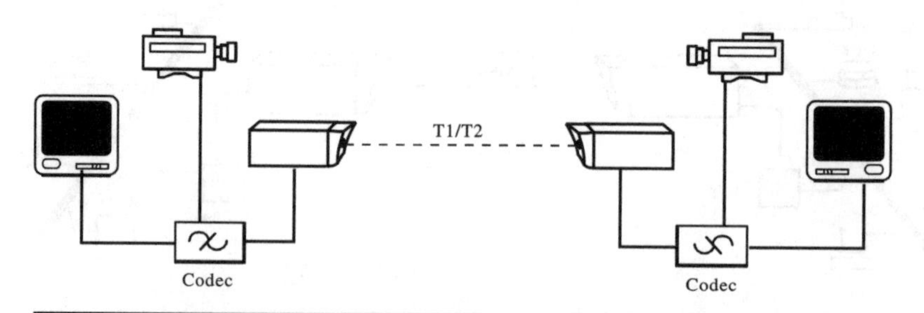

Figure 3.15 The infrared system can deliver speeds of up to 6 Mbps (T2) for video conferencing, or four links at 1.544 Mbps (T1).

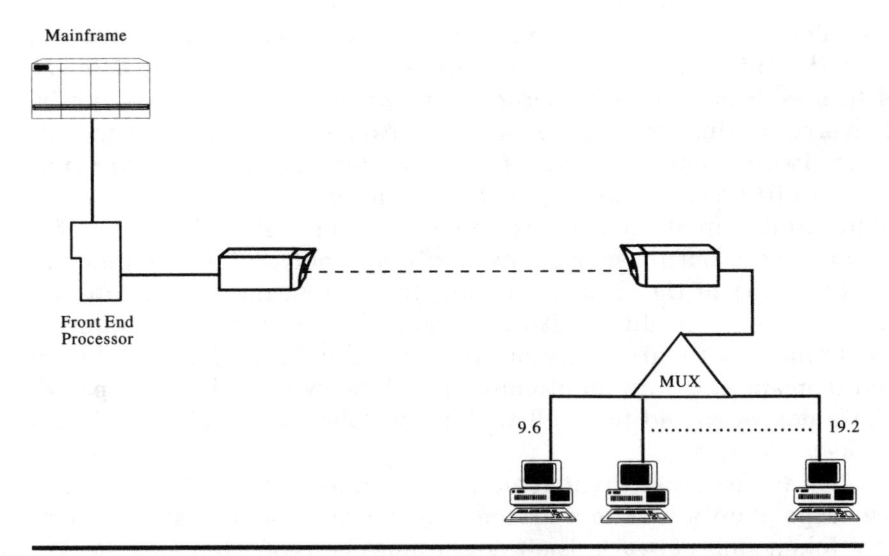

Figure 3.16 The infrared system can replace multiple DDS service for data connections to a host computer at 9.6–19.2 Kbps each.

more compressed speeds currently being offered in the industry (256,384, and 512 Kbps video services). The T1/T2 service would cost approximately $15,000 and would certainly deliver a return on investment in less than 1 year.

Dataphone Services Infrared technology can provide for high-speed digital dataphone services between a remote site and a host computer, as shown in Fig. 3.16. In this scenario, a dataphone digital service, or DDS (a trademark of AT&T Corporation), can be replaced between the two sites by using a statistical time division multiplexer (stat-mux) or a subrate multiplexer to provide several 9.6–19.2 Kbps data channels at the remote location. The system would cost approximately $10,000, yielding a payback in less than 6 months. A DDS service between these two sites at 56 Kbps could cost as much as $1800 per month.

The above examples are merely representative of the types of connections that can be provided on a 1:1 ratio. Clearly, mixing and matching the services on these infrared systems would offer even greater returns on investment.

Can an infrared system carry LAN traffic transparently?

The question always arises as to whether an infrared system can carry true LAN native speeds. The answer is yes, provided several conditions are met.

1. The system must have the bandwidth capacity to carry the native LAN traffic. This has been evidenced in the above examples when the system throughput is at least 4, 10, or 16 Mbps.

2. The system must not add significant latency to the network, since the entire LAN is based on very specific timing for the delivery of frames. Should latency become a problem, the frames will not be delivered on time, leading to discarded frames or a network timeout. The network would then require a resynchronization to recover timing. In either case, the use of the infrared system across the air would approximate 3.3 nanoseconds (ns) per meter. A 1000-meter distance between these systems would add only 3300 ns latency to the transport of the frame. Further, the system must modulate the signal onto the medium with some added delay for the equipment. Assuming a standard delay of approximately 30 ns for modulation and demodulation, an additional 60 ns latency must be anticipated. These delays should fall well within the tolerance levels of the networks as defined.

3. In an Ethernet arrangement, the connection must fall within the scope of an access to a bridge or a repeater for the cable system. In a token-ring network, the system must be connected as a station or as a multistation access unit (MAU). Once again, these connections are easily met with the use of infrared interfaces and network components.

Advantages of infrared

The advantages of infrared technology are quite numerous and should be continually considered—rather than immediately ruled out—whenever a connection is required. Many organizations automatically discount infrared as an old technology, not worthy of consideration. More astute telecommunications personnel keep it in mind as an option, at least until some other technology proves to be better suited for the job or more cost-effective. Some of the major advantages of infrared are as follows:

- Immune to radio frequency interference (RFI) and electromagnetic interference (EMI)
- Reasonably high bandwidth
- More secure technology than a radio-based technique
- Cost-effective
- Easy to install
- Efficient to move from site to site
- No license required
- Capable of traversing similar paths without interference
- Installed and running in a matter of hours

- Capable of supporting multiple standard interfaces
- Capable of handling voice, data, video, and LAN traffic either separately or on the same system, depending on the application

Disadvantages of infrared

As with any technological innovation, there is always a downside. The primary disadvantages of infrared are not as dramatic as the advantages, but they should be factored into the equation. Since the use of a light-based system is primarily determined by actual need, these disadvantages exist:

- Very limited distances, up to 2 miles (shorter distances depending on the manufacturer and the speed being used)
- Potential hazards of using a laser beam in an office environment
- Negative effects of vibration on the system
- Impaired reception of the beam because of atmospheric disturbances
- Risk of exposure to the beam from window-cleaning personnel
- Need for a clear path in a line of sight, regardless of any other condition
- Need for a stable mounting to offset sway of buildings, mounts, poles, or trees (the small footprint of the received signal allows for very little tolerance)

Fiber-Optic Systems

One of the newer products on the market utilizes a fiber-optic system of wave division multiplexing services with up to four clear channels of pulse frequency modulation video or digital data signals per wavelength window via one multimode fiber. Although this is not a wireless technique, it has come to be compared with *free-space optics.* In the conceptual model of this new service, the bandwidth of a fiber-optic channel capacity is created across free airwaves. (Hence, it is "free space," as opposed to inhibited by a piece of glass cable.) A standard fiber-based system uses different wavelengths to carry the channels of communication across a glass fiber. By contrast, the free-space optics system uses a different color of light in a different wavelength to carry various streams of voice, data, or video transmission over short distances. Obviously, since this concept uses laser beams in free space, low-power output devices are needed. Distances of 3 to 5 miles are being discussed at this time. Potential applications for free-space optics include, but are not limited to:

Security and surveillance systems

Multimedia learning over a short distance

Video multiplexing using the typical U.S. and international video standards accepted and known worldwide

Digital data at native speeds

Digital voice transmission

The low-power output is required to prevent any damage when the laser is operating in closed-in areas, such as major metropolitan centers. Also, free-space optics will have to meet all the requirements of infrared technology, including keeping the system above pedestrian traffic. However, since the visible light spectrum is involved, the signal will be less impeded by the various atmospheric disturbances that affect infrared systems.

Like infrared models, free-space systems will be fully modular and easily transportable, with the flexibility to set up and operate a transmission quickly. Since light beams will travel across the airwaves, the licensing issue will not come into play. The bandwidth will range from approximately 2 MHz to 40 MHz, yielding data rates of up to 16 Mbps for LAN traffic native at speed. Wavelengths will operate at 750, 780, 810, and 840 nm. This new communications medium will obviously offer some very attractive options to end users for access to the long-distance carrier's point of presence (POP) and ultimate access to the long-distance networks. Local, campus, metropolitan, and wide area networks will all have some use of free-space optics. Consider such a system as an alternative to accessing the long-distance network, or as a means of bypassing the local loop in the telephone company arena.

Chapter

4

Cellular Communications

Evolution of Cellular Technology

As early as 1946, the first commercially available radio-to-telephone service was introduced in St. Louis, Missouri. Telephone company subscribers in that city were having problems getting a dial tone from the central office. Subscribers who were lucky enough to get the dial tone then experienced problems with call completions. The use of radio transmitters linked to the telephone company aided in basic telephony. However, the major problem was that the radio telephone network consisted of a limited number of frequencies (channels). The system was mounted atop a large building. All calls within the city had to be routed through this centralized tower because of the limited number of channels. Twelve channels were available in the St. Louis network. Each channel was used on a high-powered radio transmitter to provide the required coverage. Further, the system operated on a one-way (simplex) basis. Only one party could speak at a time. This was analogous to the holder two-way "push to talk" radio technology. The system took quite a bit of getting used to because of the unidirectional transmission.

By the mid-1960s mobile telephone technology had undergone some changes. An improved mobile telephone service (IMTS) offered users direct access to the dial telephone network. Also, the need to "push to talk" was eliminated. Some of the enhancements were accomplished by a higher-powered radio system (operating at 200 watts or more of output). Two separate channels provided duplex operation. Coverage for the mobile phone was approximately 20–25 miles, but interference on radio systems could occur at distances up to 100 miles. The result was limited availability of channel capacity and frequency reuse.

TABLE 4.1 Limitations and Strengths of IMTS

Limitations	Strengths
Limited service areas	No push-to-talk procedure
Poor transmission	Two-way transmission
Excessive delays in call setup	Separate frequencies for transmit and receive
Limited frequency reuse	
Demand in excess of capacity	

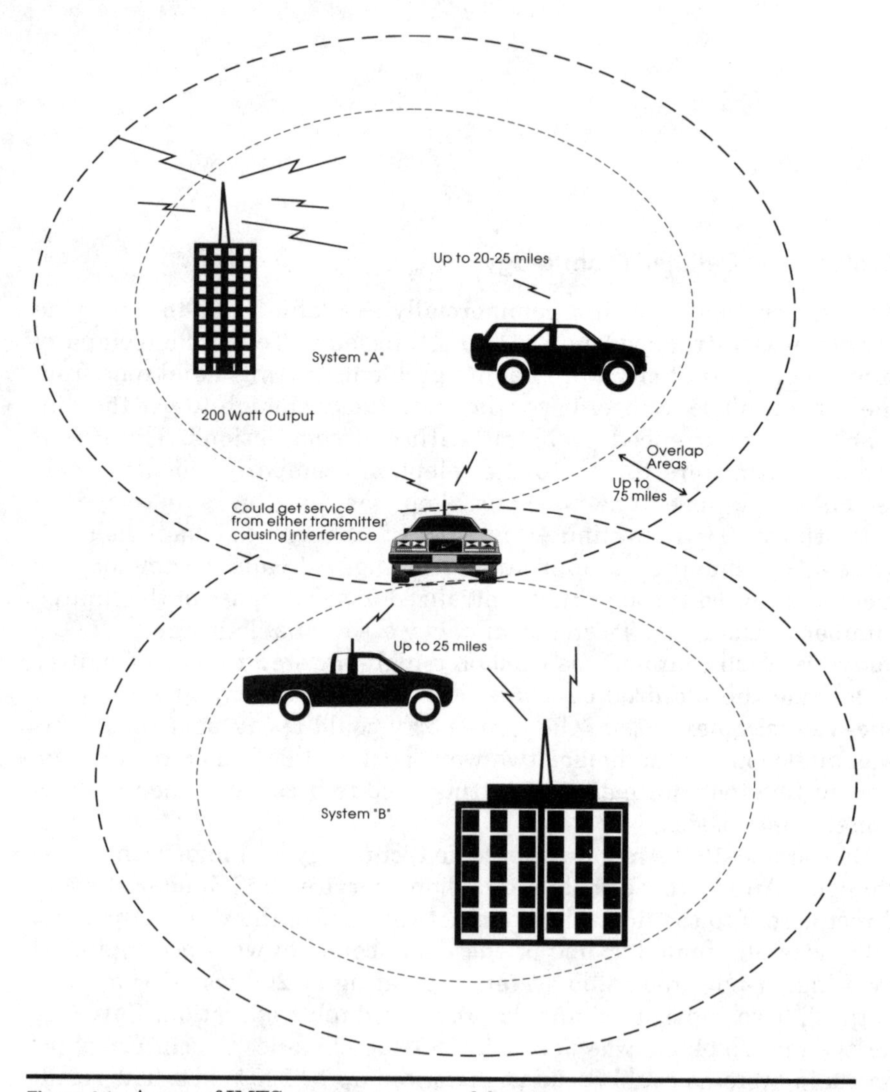

Figure 4.1 A map of IMTS coverage areas and frequency reuse.

Table 4.1 summarizes the limitations and strengths of the IMTS system. A map of IMTS coverage areas and frequency reuse is presented in Fig. 4.1.

Traditional mobile systems were designed around very high power outputs to cover large geographical areas. Therefore, the ability to use IMTS and its cousin AMPS (advanced mobile phone service) was governed by the radio frequency spectrum and distance limitations of the outputs. If an attempt was made to use frequencies that were too close to one another, interference would result. It goes without saying that the radio frequency spectrum is a finite and limited resource which is rapidly becoming constrained.

Both the IMTS and AMPS systems operated in different frequency bands (44 channels were assigned to public mobile services operating in 35–44 MHz, 152–158 MHz, and 454–512 MHz). Service was basically high-output line of sight over broad geographical areas, often jamming neighboring frequencies out of the radio spectrum. Therefore, frequencies could be reused over an area of only 50–100 miles. Because of the 50–100 mile separation (buffer zone), frequencies were limited. Normal mobile operations consisted of high-output transmitters operating at 125–250 watts of output in order to maximize coverage, transmission strength, and clarity. The systems operated by selecting the first available channel for transmission.

As mobile phone users gained sophistication, the need to communicate from anywhere, at any time, became more urgent. Even though the mobile telephone network was originally a voice-only service, the demand was quite high. Unfortunately, because of the high output capabilities and limited frequencies available, waiting lists with lead times as long as 2 years were not uncommon. Only the very affluent could afford these systems. Further, regardless of affluence most radio-based telephones were assigned to "life and limb" services (fire, police, ambulance, etc.). This led to disparities in the industry across the country.

Around the same time as the AT&T divestiture (the breakup of "Ma Bell" from the local telephone companies), developers of radio telephone interconnection were experimenting with new "cellular" communications. In actuality, cellular communications were designed and ready to go in 1974! The 800–900 MHz portion of the ultra high frequency (UHF) band was allocated for cellular use. After various delays and hearings, the service was finally demonstrated in Chicago, in 1978.

It wasn't until 1981 that the FCC finally set aside 666 radio channels for cellular use in the United States. These frequencies were assigned to two separate carriers. The lower frequencies were reserved for "wireline companies"—the regulated providers (local telcos). The higher frequencies were reserved for "nonwireline" carriers—the competitors to the telcos. Both groups of carriers (wireline and nonwireline) are licensed to operate in a specific metropolitan service area (MSA) or rural service area (RSA).

Dynamics of Cellular Transmission

Cellular technology overcomes the limitations of conventional mobile telephone systems. Areas of coverage are divided into small honeycomb (hexagonal) cells that overlap at the outer boundaries (see Fig. 4.2). Frequencies can be divided into bands or cells with a protection zone established to prevent interference and jamming of the neighboring cell's frequencies. The cellular system uses much less power output for transmitting. The transmitter is designed around 3 watts of output at the radio, so that frequencies can be reused much more often and closer to one another. The average cell is 3–5 miles across; 2–10 miles are possible, with the distance

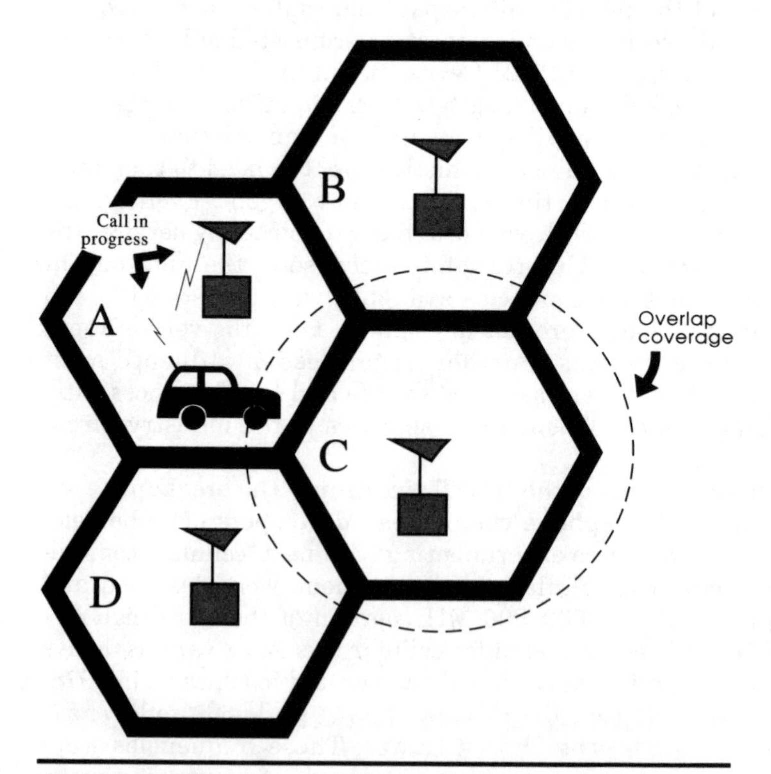

Figure 4.2 A call in progress will be handled by the cell in which the user is located.

depending on the expected number of users in a geographical area. The greater the number of users, the closer the transmitters. In rural areas, the cells are much farther apart. Since the power output can be reduced to 3 watts for mobile phones, and .6 watts or less for portable cellular sets, interference is limited. The FCC reserves half the frequencies for the wireline companies and half for the nonwireline services. Each carrier uses approximately 312 frequencies for voice and data communication and 21 frequencies for control channels (see below).

Controlled supervision and switching of calls is particularly critical in a mobile environment. Without the dynamic switching and control necessary to facilitate a seamless shift as the vehicle moves from one cell to another, all communication could be terminated. In conventional mobile systems, the radiophone would degrade as the vehicle moved away from the base station, frustrating both the caller and the receiving party. As the user went beyond an area of coverage, the call would be cut off, further frustrating both parties. With cellular communications, when a call is in progress and the caller moves away from the cell site toward a new cell, the call gets "handed off" from one cell to another. Here's how the process works.

Log-on

Each cellular phone is given a unique identity, or numeric assignment module (NAM). The cellular phone is also assigned a home area for traffic, so that messages can be sent to the phone through separate control channels in the home area. Whenever the phone is turned on, a check is sent across the control channel to verify that it is operating within its home area. A phone operating outside the home area will be required to reregister, notifying the home unit of its location. As a mobile unit moves from one area to another, it must continually send messages to the mobile telephone switching office (MTSO) to verify its location. All new calls are directed to the new traffic area. In Fig. 4.3 the set is turned on, causing the registration message to take place.

Monitoring

After the power-on sequence, the cellular phone monitors the dedicated control channels to get information on local paging channels. It tunes itself to a suitable available paging channel and goes into an idle state. In this idle state it listens to the data being transmitted on

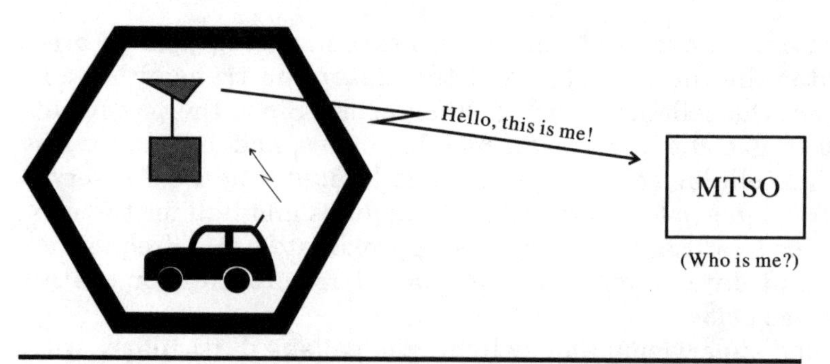

Figure 4.3 When the unit is turned on, a control message is sent to verify location.

the channels. In the event that the signal level falls, the unit will go back into a scanning mode until it finds a new paging channel. In Fig. 4.4 the cellular set is constantly listening for the control information on the paging channel. In Fig. 4.5 the sequence starts over as the cellular unit rolls into another cell.

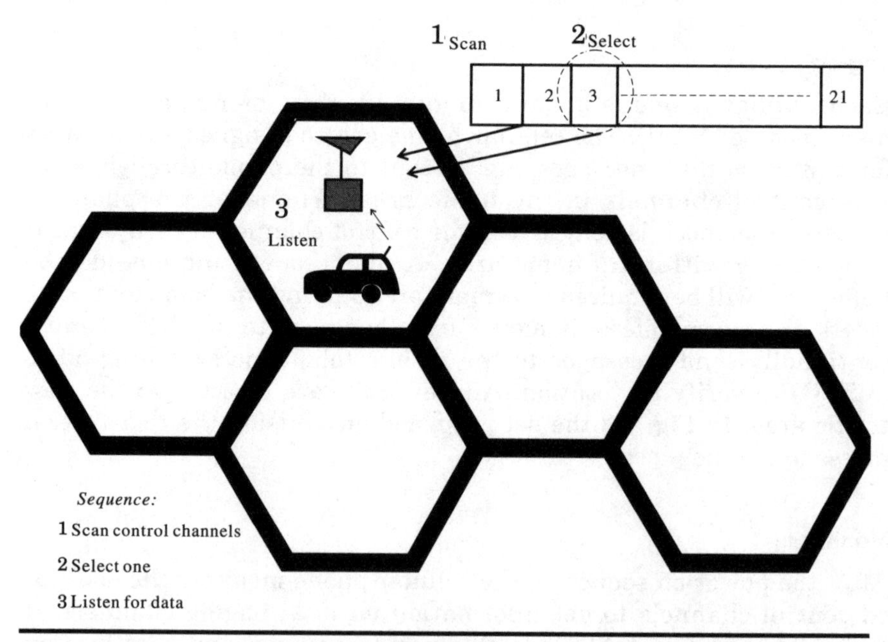

Figure 4.4 Once powered up, the cellular set selects a paging channel and listens to it.

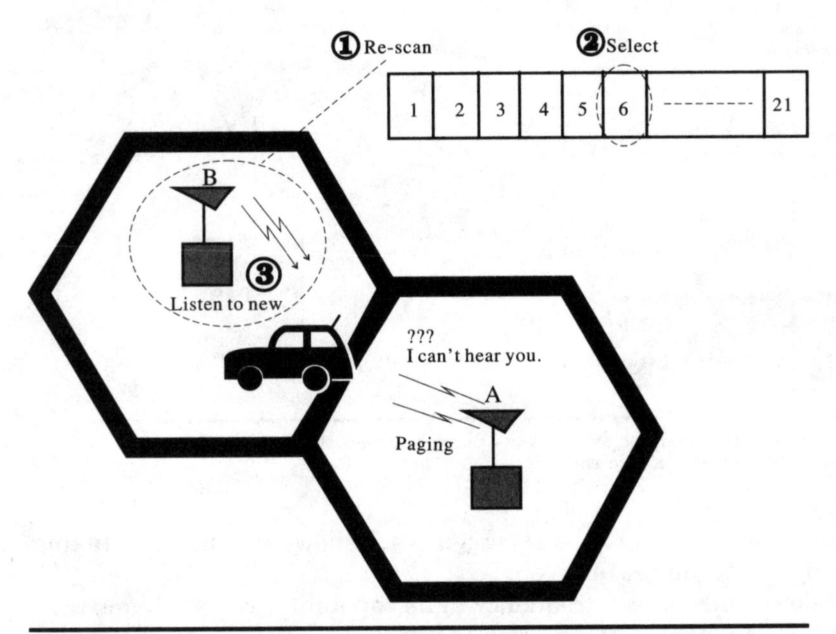

Figure 4.5 When the signal strength falls or a new cell is entered, the cellular set reinitializes the sequence.

Outgoing calls

To make a call, the user enters the telephone number through the keypad into the set's memory and then presses the "send" button. The cellular set accesses the system by scanning the available access channels (located by using the paging channels). Once the set locates an access channel, it transmits a request to set up a call. It waits for a reply from the system on the access channel. The reply comes back in the form of an instruction to use a specific channel, causing the set to tune to the frequency of that channel. Figure 4.6 shows this process as a dialogue between the cellular set and the system.

Incoming calls

When an incoming call is routed to a cellular phone, the cellular system first has to find the set. This is done through the paging channel. A page is sent out from the system to all cells in the set's traffic area. When the set receives the page, it responds to the system. The system will instruct the set to use a specific channel to receive the call. The set then retunes itself to the specific frequency, ready to talk. The

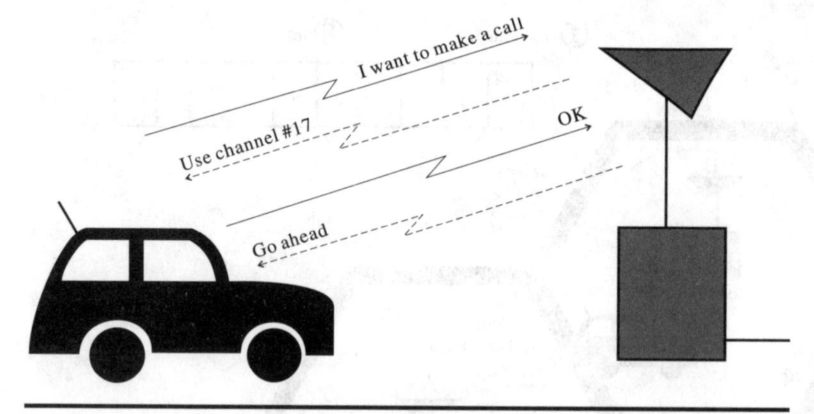

Figure 4.6 A handshake, or sorts, takes place between the sender and the system to establish a connection.

system sends the call to the set. Figure 4.7 shows the dialogue taking place for an incoming call.

Mobile phones have a frequency-agile capability, which means they can change to any of the operating frequencies in the area to use services. The phones use microprocessor logic to respond to incoming

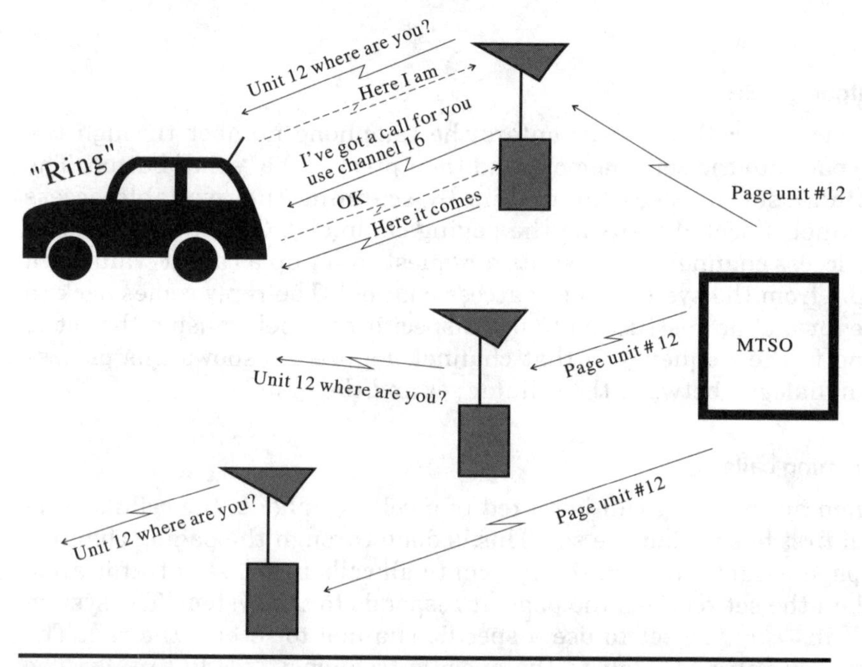

Figure 4.7 The incoming call is set up through a dialogue between the system and the cellular set.

calls and shift to various frequencies to receive and make calls. At least one channel is designated for calling and control in each cell. As a call comes in, the cell site controller assigns a channel and then directs a frequency synthesizer inside the mobile unit to shift to the appropriate frequency.

Handoff

Once a call is in progress and the switch from cell to cell becomes necessary to continue the call, a handoff takes place. As the cellular phone approaches an imaginary line on the earth, the signal strength transmitted back to the cell site starts to fall. The cell site equipment sends a form of distress message to the MTSO, stating that the signal is getting weaker. The MTSO then orchestrates the passing of the call from one cell site to another. Figure 4.8 shows the initial sequence as the cell site notifies the MTSO that something is going on.

Immediately after receiving the message from the cell site, the MTSO sends out a broadcast to other cell sites in the area. Its request is to determine which site is receiving the cellular user's signal the strongest. Each site responds accordingly. In Fig. 4.9 the MTSO's initial broadcast goes out across the network.

As the responses come back to the MTSO, the particular cell receiving the cellular user's signal the strongest is selected to accept the call. The MTSO directs the receiving cell site to set up a voice path parallel to the site losing the signal. When the setup is ready, the MTSO sends a message to the cellular set to change to a new frequen-

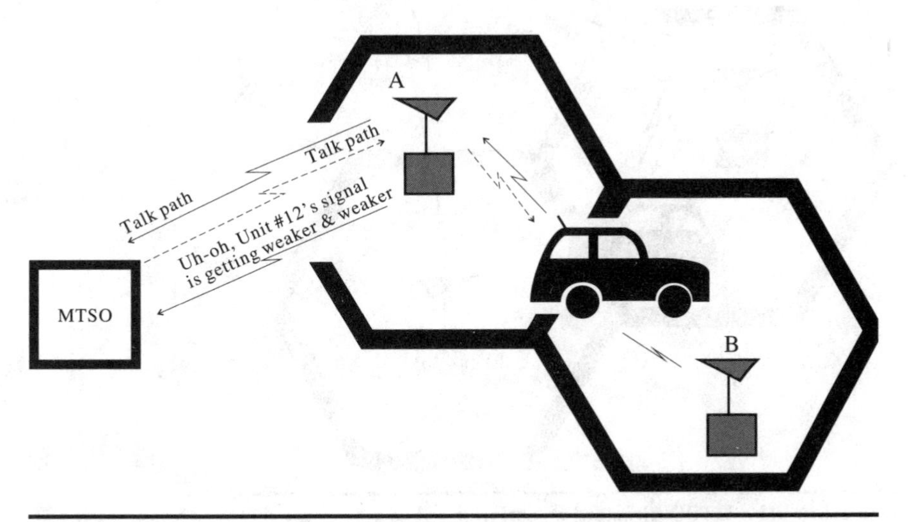

Figure 4.8 The cell site senses the drop in signal strength and alerts the MTSO.

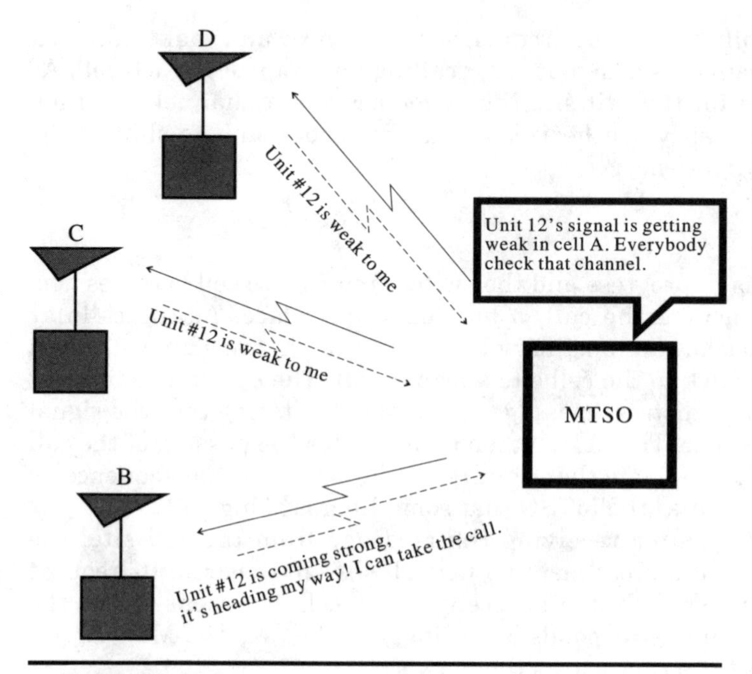

Figure 4.9 The MTSO begins its initial sequence to prepare for the handoff.

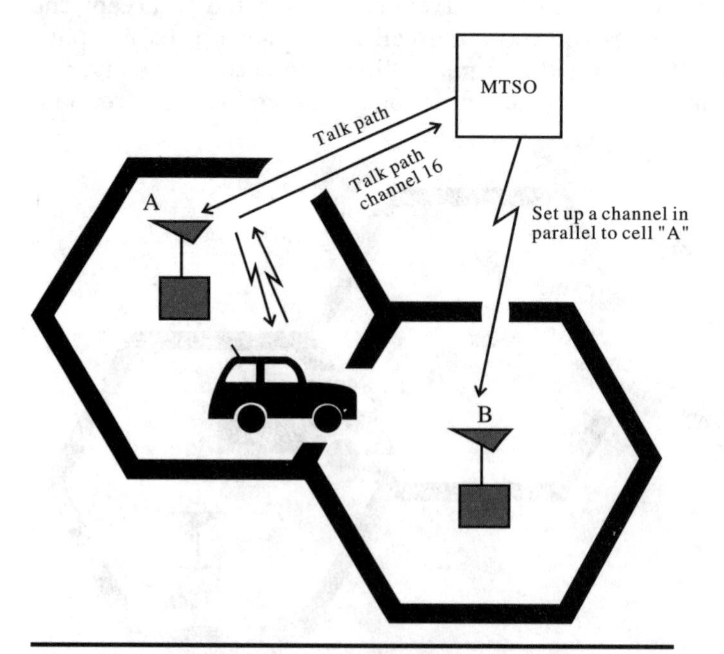

Figure 4.10 The MTSO directs the receiving cell to set up a channel.

cy (resynthesize) and get ready. The cellular set then retunes itself to the new frequency assigned to the receiving cell site, and the handoff takes place. The process takes approximately 100 milliseconds.

Figure 4.10 shows the first stage of the handoff. The cell site then advises the MTSO that the parallel channel has been set up, as shown in Fig. 4.11. Immediately the MTSO sends out a control message informing the cellular set of its new frequency, as shown in Fig. 4.12. At this point the handoff takes place, as shown in Fig. 4.13, and the original channel becomes idle.

Cellular Components

The various components of a cellular system merit investigation in greater detail. The three main hardware components of the cellular network are:

Cell site (base station)

Mobile telephone switching office

Cellular (mobile) handset

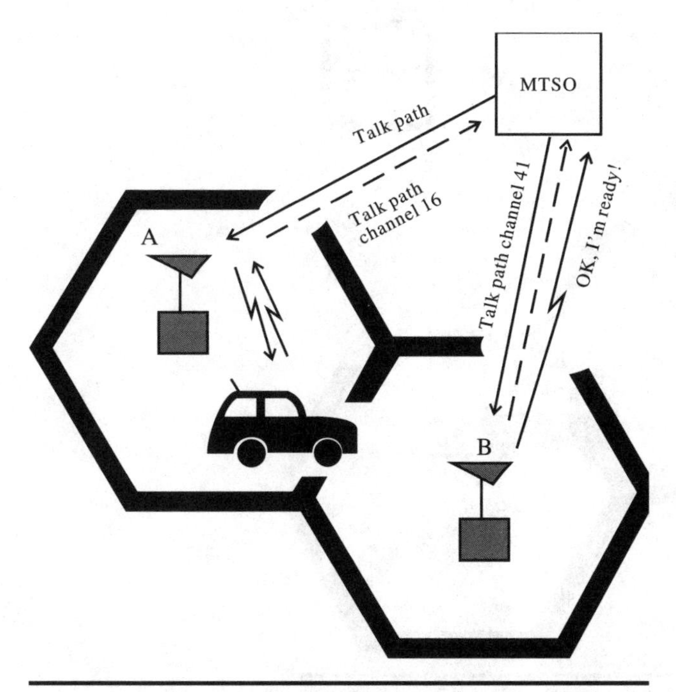

Figure 4.11 Once the path is set up, the cell site alerts the MTSO.

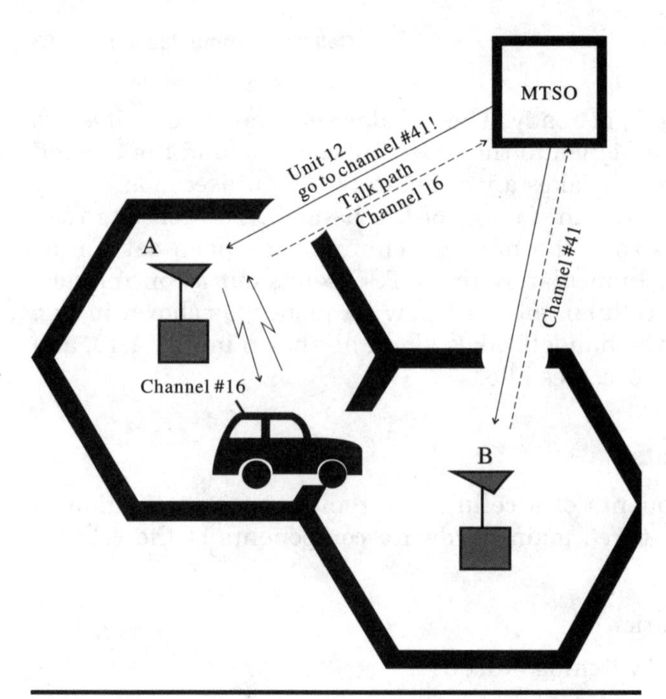

Figure 4.12 The alert goes out to the cellular site advising it to go to a new channel.

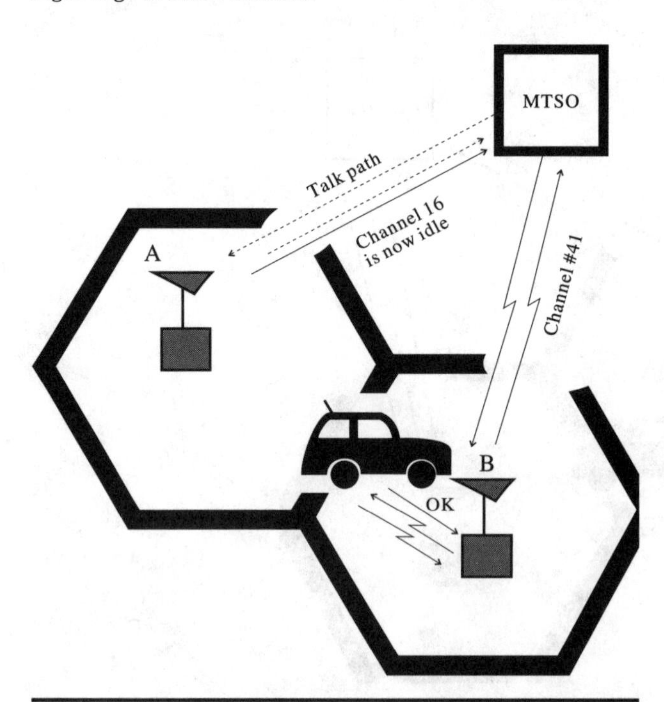

Figure 4.13 The unit returns to the channel assigned and the handoff takes place.

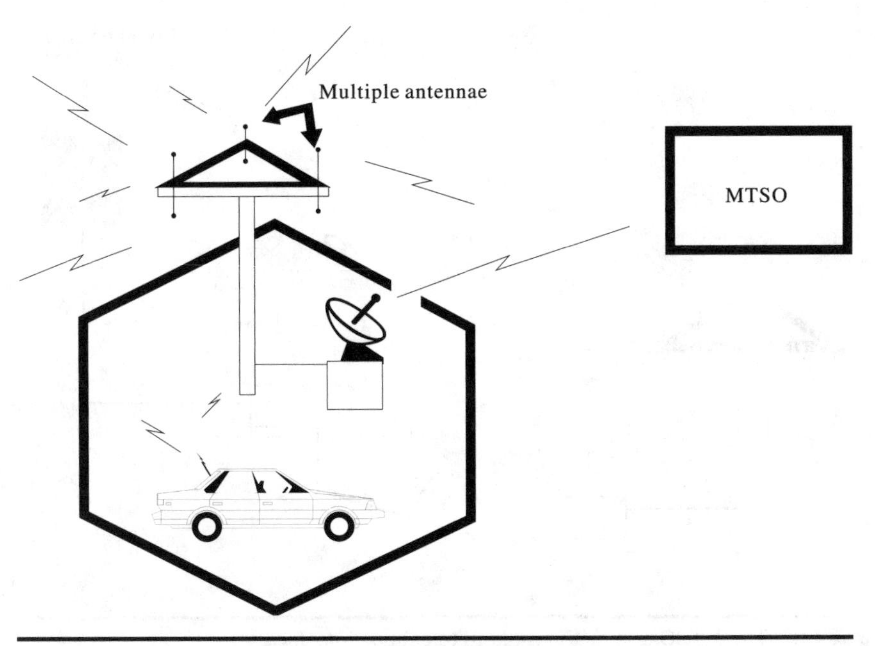

Figure 4.14 The cell site acts as the user-to-MTSO interface.

Cell site

The cell site consists of a transmitter and two receivers per channel, a controller, an antenna system, and data links to the cellular office. Directional antennas are used to cover the area within the cell, an area approximately 5 miles across. Handoffs from antenna to antenna take place at the cell. As many as 128 channels can operate within a cell, depending on the area of coverage to be provided. Most systems operate with much fewer channels—45 pairs of frequencies are typical. As cells get congested, a subsplitting or sectioning technique can be used to accommodate more users. The cell site operates as the user-to-MTSO interface, as shown in Fig. 4.14.

Mobile telephone switching office

The MTSO is the radio equivalent of a Class 5 end office in the normal telephone system hierarchy. The MTSO is the physical provider of connections from the cellular radio through the cell site to the local exchange carrier. The MTSO has dedicated circuits between the cell site and its logic, using either a landline (wired) phone or a mobile (microwave) unit at the four-wire E&M level. E&M signaling uses a separate signal path (wire) for transmitted signals (M) or received sig-

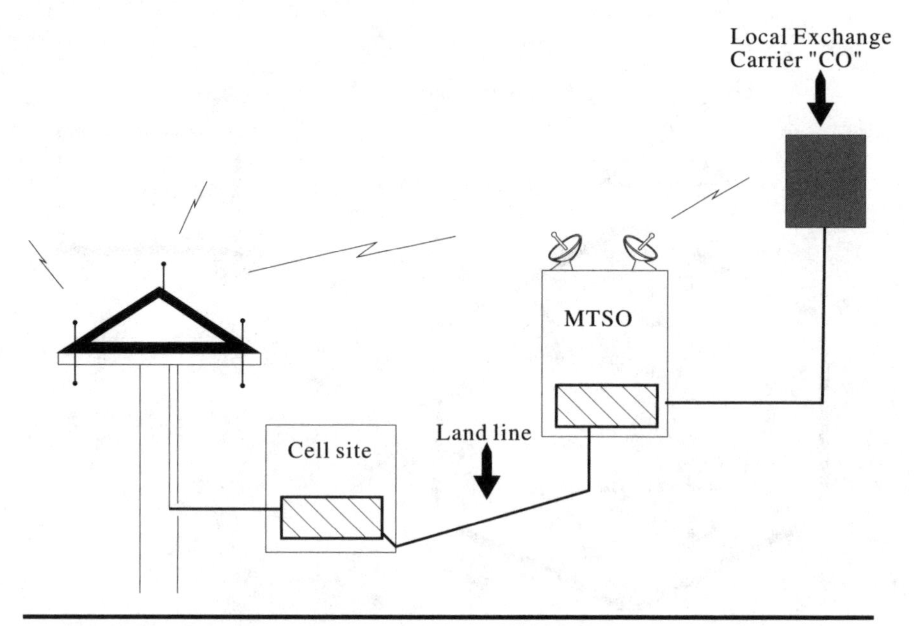

Figure 4.15 The MTSO acts as the radio-to-telephone interface.

nals on its own wire (E), referred to as "ear & mouth." The control portion of the network uses X.25 packets to establish, set up, release calls, and so on. In Fig. 4.15 the MTSO connects the cell set through the cell site and on to the local telephone company office. The connection can be on either a landline or a microwave radio system between the points.

Cellular (mobile) handset

The handset houses a transceiver capable of tuning to all channels within an area. These are frequency-agile units capable of receiving and transmitting on all 666 frequencies, as opposed to the fixed-frequency units of old. The components of the mobile set are shown in Fig. 4.16. The transmitter, receiver, and synthesizer are the intelligent portions of the mobile set. The diplexer allows the set to use a single antenna for the transmit and receive portions of the cellular communication. The logic unit and numeric assignment module are chipsets housed inside the mobile set.

The major components of the mobile unit are:

Handset

Number assignment module (NAM)—an electronic fingerprint (32-bit binary sequence)

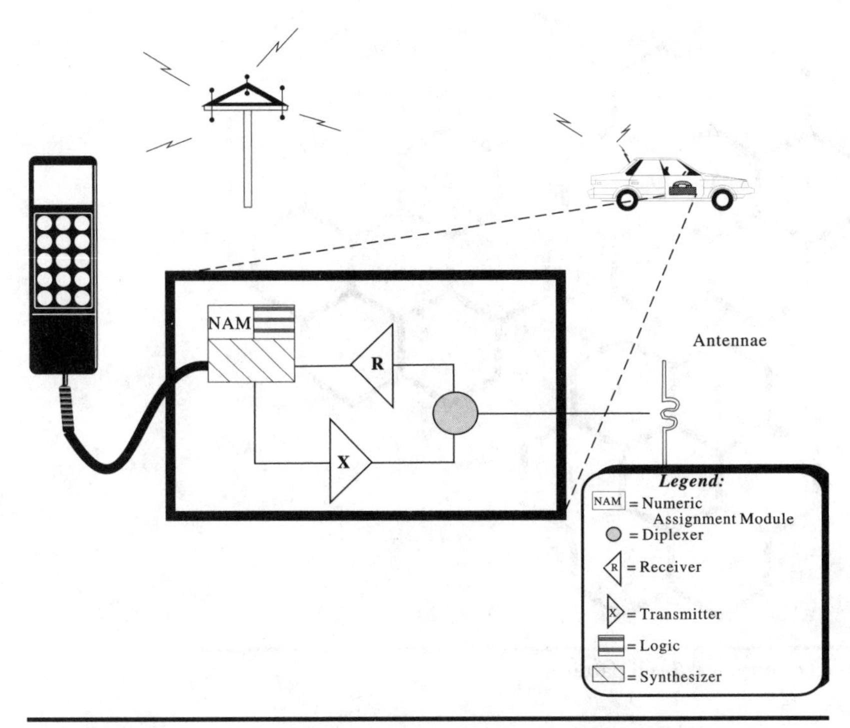

Figure 4.16 The cellular unit sets up the calls to the cell site, using a frequency-agile capacity to synthesize the necessary frequency.

Logic unit

Transmitter

Receiver

Frequency synthesizer to generate frequencies under control of the logic unit

Diplexer to separate transmitter and receiver functions

Antenna

Cellular Coverage and Channels

As mentioned above, cellular technology has become effective through the use of smaller areas of coverage and lower-powered output devices. The basic cellular pattern is to use channels of radio frequency in clusters (called cells). Since the cells are small and the output is limited (3 watts of output on a vehicular set, or .3–.6 watts of output on a handheld device), the frequencies can be used again and again. This allows far better management of the limited radio frequency

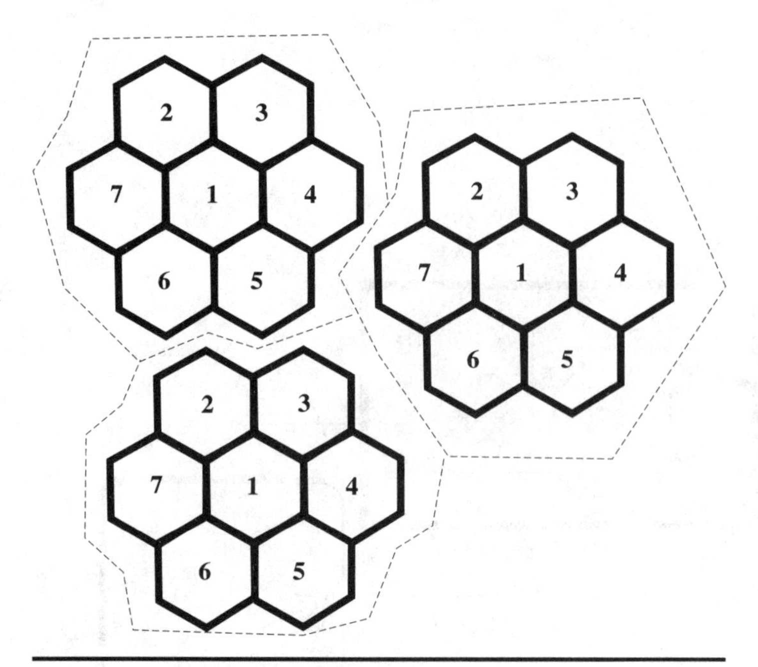

Figure 4.17 A typical 7-cell pattern.

spectrum. The separate cell clusters create minimum interference, since the frequency reuse must be at least two cells away. The patterns are grouped and normally use 4, 7, 12, or 21 cells to allow for adequate coverage and to prevent cross talk and interference. Figure 4.17 is a typical 7-cell pattern showing the repeated use of channels and the separation between reuse.

Overlapping coverage

Each cell has its own cell site (radio equipment) with an overlap into adjoining cells. This allows for the monitoring of the adjacent cells to ensure complete coverage. The cells can sense the signal strength of the mobile and handheld units in their own areas and in the overlap areas of each adjoining cell. This is what makes the handoff and coverage areas work together. Figure 4.18 shows the overlap coverage in the 7-cell pattern described above.

Channel functions

A cellular radio system provides two types of channels. The first is a duplex control channel used to transfer information on call setup and breakdown. The second is a duplex voice channel used to carry the two-way voice conversation. The control channel is broken down into

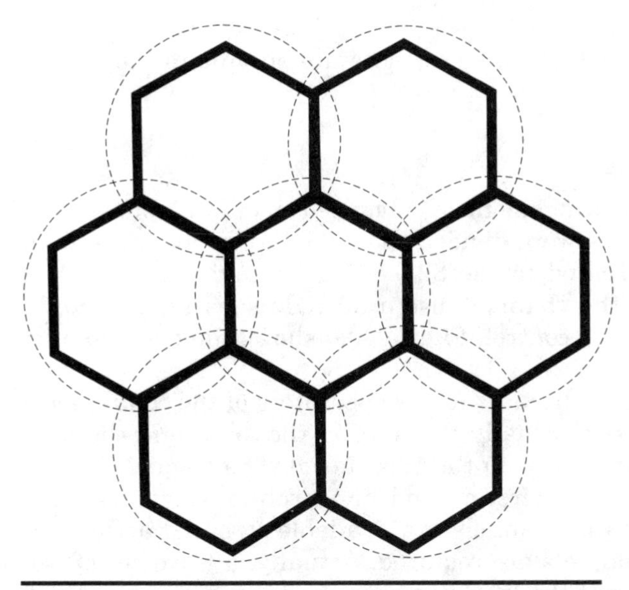

Figure 4.18 The cells overlap to provide greater coverage.

three dedicated functions: control, paging, and access. These functions are provided in various implementations to meet the carrier's use and traffic demands.

As soon as the user set is turned on, it scans these dedicated channels, which are usually programmed into the set's memory. Immediately the set tunes to and locks in on the strongest channel. It then listens for information passing along this channel, such as which channels are allocated to paging. Once the set locates a paging channel, it receives information regarding call traffic in the area and other network-specific instructions. At this point the set goes into an idle state and monitors the messages being transferred on the paging channel.

Figure 4.19 shows the cellular spectrum for voice, control, access, and paging channels. Note that the frequencies are plotted but not

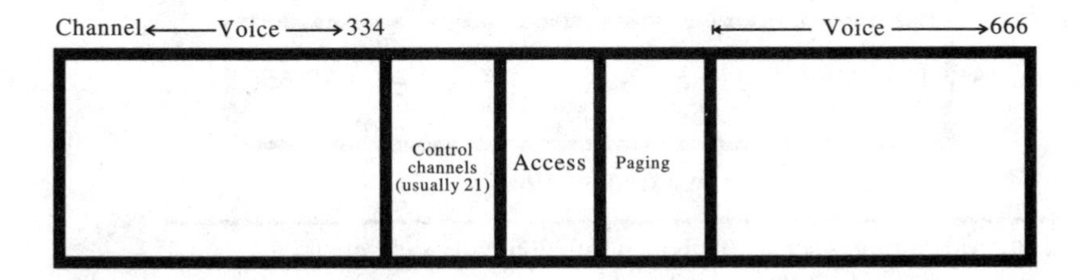

Figure 4.19 The arrangement of channel types in the overall spectrum.

the exact numbers of channels, since these are specific to the vendor and carrier.

Allocation of frequencies

In the first cellular arrangement of advanced mobile phone service (AMPS), the frequencies were divided into 666 duplex channels, with frequency ranges allocated in the 825–845 MHz and 870–890 MHz bands. In each band the channels use a 30 KHz separation, and 21 channels are allocated to control. Figure 4.20 shows the channel allocation.

The FCC has approved licenses for two operators of the cellular service: the wireline carrier (usually the telco in the area) and the nonwireline carrier (a competitor to the local telco). The frequencies are split equally between the wireline and nonwireline operators. This means that only half the channels are available to each carrier and two sets of control channels are required. Actually, a third set of frequencies is available, but the FCC has not yet assigned any to a third carrier. Since cellular service was the first attempt to provide competition, it was decided to use only two carriers to start.

Four signaling paths in the cellular network provide for signaling and control as well as voice conversation. These can be broken into two basic functions: (1) call setup and breakdown and (2) call man-

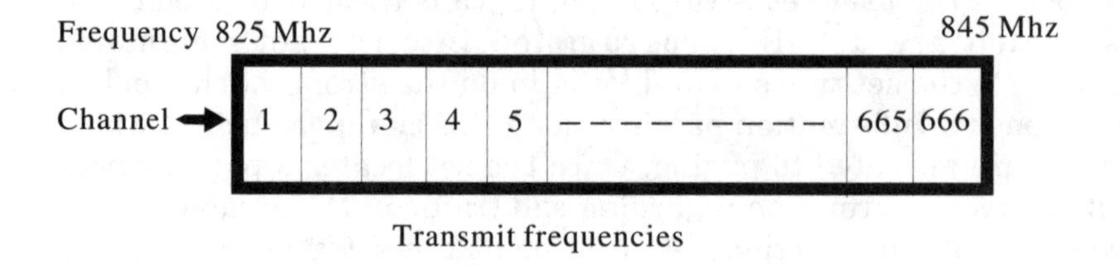

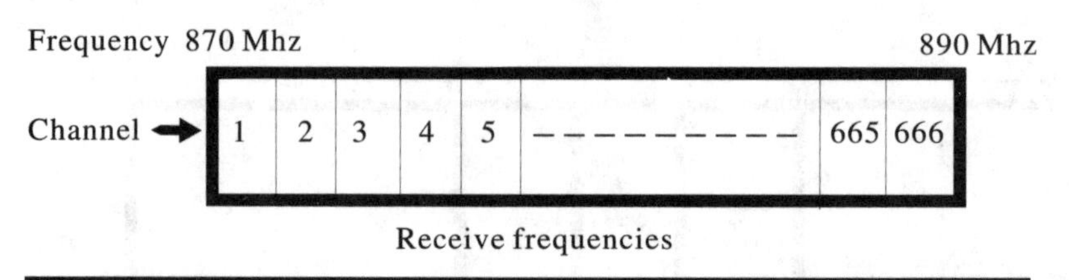

Figure 4.20 AMPS frequencies are divided into 666 duplex channels.

TABLE 4.2 Directionalized Control

Function	Channel
Call Setup	Forward control
Call Setup	Reverse control
Call management and conversation	Forward voice
Call management and conversation	Reverse voice

agement and conversation. In each of these two functions, forward and reverse channels provide the directionalized flow of information, as shown in Table 4.2. The forward control and reverse control channels are used to set up and break down calls and to manage the cellular phones in the system. These channels are used strictly for signaling and control; they do not carry conversations.

The forward voice and reverse voice channels are used to manage the actual calls. Data are transmitted on these channels before, during, and after a call. To prevent confusion to the user during a conversation, a meeting of the speech path occurs while management information (data) is transmitted. Voice, data, and supervisory information are all sent over the network, each modulated differently to provide distinctive information. Figure 4.21 shows the four channels used at each cellular set.

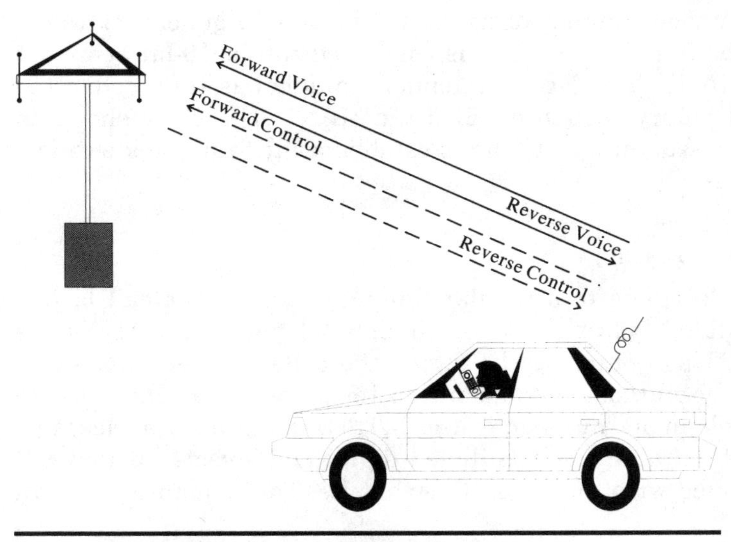

Figure 4.21 Four channels are used at each cellular set.

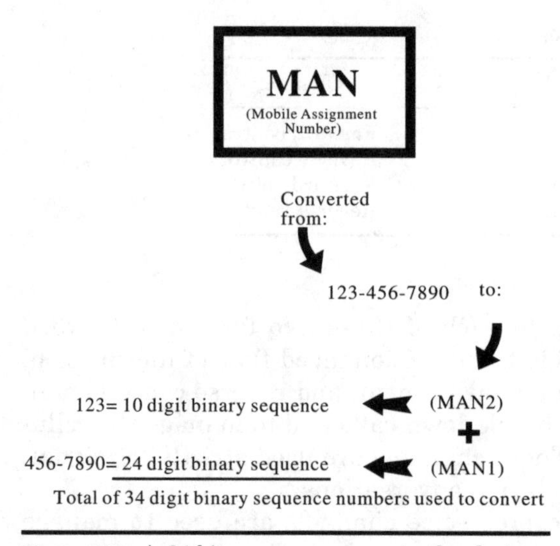

Figure 4.22 A 34-bit sequence converts the decimal number.

Routing Cellular Calls

When a cellular set is turned on, it reads a certain amount of information from its memory. This information is normally programmed in by the carrier to give the user a unique mobile assignment number (MAN). The MAN is a 34-bit binary number derived from the 10-digit decimal telephone number assigned to the set. In general it works this way: The 3-digit area code is converted into a 10-bit sequence called the MAN2. The 7-digit telephone number is then converted into a 24-bit binary sequence called the MAN1. These are shown in Fig. 4.22. The sequence of 10-digits establishes the unique identification for the set.

Mobile-to-landline calling

From a mobile phone to a regular (landline) phone, the call begins when the caller keys in the 7- or 10-digit telephone number of the called party. Upon entering the digits, the caller presses the "send" button and the number is transmitted to the cell site. The cell site relays the telephone number to the MTSO, using a data message. The MTSO then analyzes the dialed digits and selects an idle cell trunk associated with the requesting cell site. The sequence is shown in Fig. 4.23.

The MTSO concurrently processes the call to the local exchange

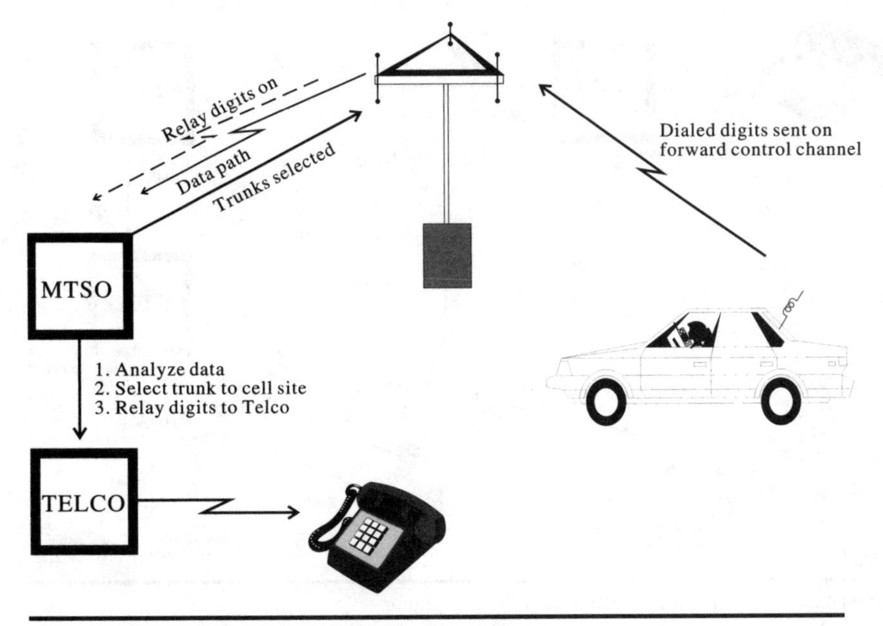

Figure 4.23 A call from a mobile to a landline phone.

office and repeats the dialed digits to the telco. The call is then handled as a regular switched call in the telephone hierarchy.

Landline-to-mobile calling

From a wired telephone the local exchange office pulses out the cellular number called to the MTSO over a special trunk connecting the telco to the MTSO. The MTSO then analyzes the number called and sends a data-link message to all paging cell sites to locate the unit called. When the cellular unit recognizes the page, it sends a message to the nearest cell site. This cell site then sends a data-link message notifying the MTSO that the unit has been found and identifying the cell site that will handle the call.

The MTSO next selects a cell site trunk connected to that cell and sets up a network path between the cell site and the originating trunk carrying the call. The process is shown in Fig. 4.24.

Cellular-to-cellular calling

The easiest call an MTSO can route is from one cellular unit to another in the same service area. The MTSO receives the dialed dig-

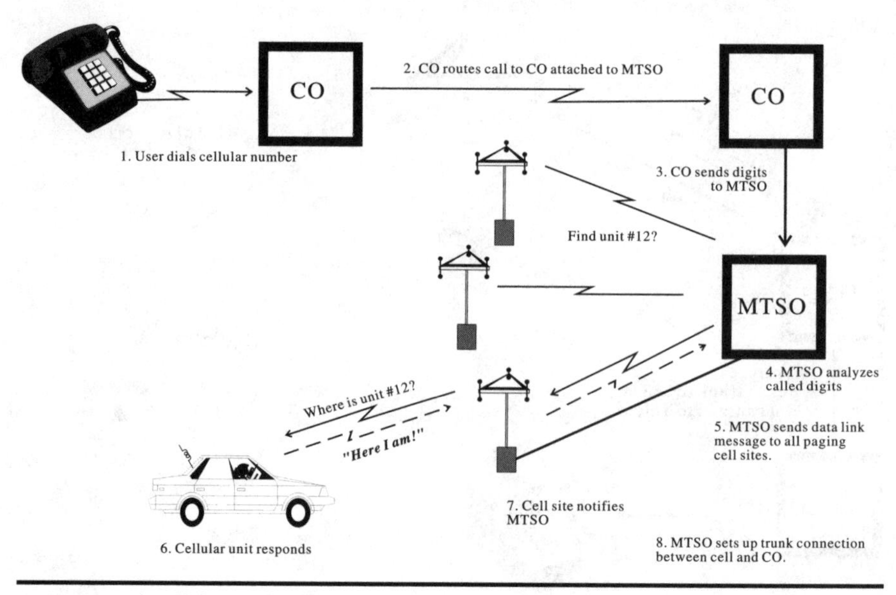

Figure 4.24 A call from a landline to a mobile phone.

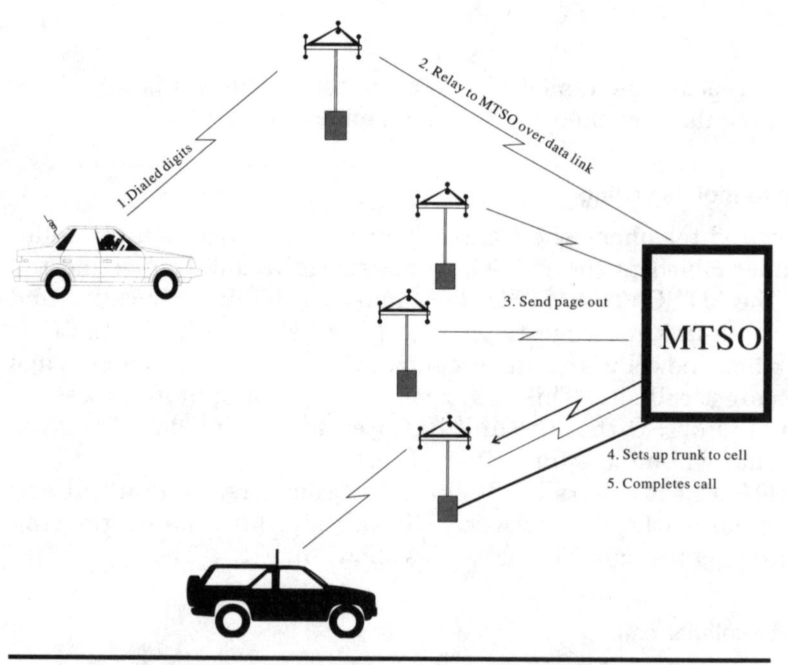

Figure 4.25 The cellular-to-cellular call in the same area is the easiest to complete.

its from the calling phone, determines that it is another cellular phone, then routes the call to the receiving cellular set through the appropriate cell site. No interconnection is needed with the local telco switching offices. Figure 4.25 shows how this call is processed through the cellular network.

Chapter

5

Wireless Personal Communications

Probably no single development has done more for wireless technologies than cellular communications. The availability of frequency reuse, reduced power outputs, and lower up-front costs to introduce customers to the service have all had a significant impact on wireless telephony. To establish a cellular operation in an average-size city would cost approximately $500 million in up-front carrier costs alone. Start-up costs are considerably higher in large metropolitan areas. Adding new customers and maintaining older ones has proved to be a formidable task for even the strongest players in the market. Until recently, the major pitfalls in the cellular market have been:

- High up-front costs for carriers
- Higher-than-average cost per minute for users
- Expensive equipment (with a cell site costing around $850,000 and a cellular phone originally costing around $3000)
- Uneducated users who do not understand the costs associated with cellular communications and the concept of paying for incoming airtime
- Risk of fraudulent use of the cellular user's account
- Theft of airtime, sets, and information

Defining Personal Communications

It is little wonder that early cellular operators added new users at the rate of 150,000 accounts per month, but lost 100,000 at the same time because of "sticker shock." Now a newer technique is emerging, one that has been given various labels by various sources: personal communications services (PCS) by the FCC, personal communications networks (PCN) by the rest of the world, and even personal communications interface (PCI) by a recent entrant in the field, Northern Telecomm Inc.

Whatever the label, this new service is geared toward the masses. Why should the individual telephone user be tied to a pair of wires or to a fixed location with an associated number when network suppliers can offer the user a tie through an intelligent network capable of finding a called party at any location? Therefore, the concept of a wireless local loop providing basic telephony services to small businesses and residential consumers was created. This concept appeals to the mass market at a time when the population is becoming far more mobile. Cordless telephone technology has already made its niche in the industry; a logical extension of this service is to provide unrestricted access and use regardless of where the recipient of the call is at any time.

Consider the possibility of a cordless telephony capability anywhere in the world, and personal communications becomes far more than a myth. Throughout the evolution of wireless communications, users have wanted the ability to be available at any time. In the past, this was accomplished through a personal pager or a two-way radio. Unfortunately, these systems were always limited to a specific area, such as a major metropolitan center and its surroundings (typically a 40- to 50-mile radius from the user's home base). This distance limitation led to complex gyrations in the use of paging and radio handheld systems.

At first, the industry responded with a more robust application through satellite and radio repeaters so that users could move out of the home area and be reached virtually anywhere in the country. Pockets of "dead space" remained, but these were somewhat limited. Even when two-way improved mobile telephone service (IMTS) was deployed, several limitations persisted. Limited frequencies, long waiting lists, and expensive service all kept the use of this technology to a minimum. Cellular telephony improved the situation, but the cost of airtime was considered prohibitive. In a period when airtime costs should have been dropping, carriers and providers of cellular service raised their rates to compensate for losses stemming from high churn (turnover) among customers and from a new phenomenon known as cellular theft or fraud. The real user was penalized because of losses being sustained by the providers.

Today the industry wants to deliver wireless communications in the form of telephony, data, paging, and E-mail applications to the mass market. To do so, suppliers need a newer concept that will allow for the rapid deployment of equipment and the more diverse coverage of all areas around the country, then the world. Rapid deployment will make major investments far less lucrative, so the proposed carriers and providers are looking for ways to introduce personal communications services to the masses with limited investments and large-scale frequency reuse. One option being considered by carriers and providers is to purchase and install microcells and picocells, as discussed later in this chapter. First and foremost, the carriers are looking to provide much smaller systems and equipment and to reuse the frequency spectrum more and more. The obvious goal is to provide the equivalent of a wireless dial tone to the masses (both businesses and residences) at an affordable price.

When personal communications technology enters the market, the industry pricing may be as high as or higher than the monthly costs for cellular communications and wired dial tone from the local telephone company. However, as the proliferation of the service takes place, the costs will begin to fall dramatically. One could ultimately expect to see this form of service renting for $10.00 per month and $.10 per minute of usage. It will take some time for the service to get established, but it should happen prior to the turn of the century. Spotty services and coverage in the early implementation stages will cause some dissatisfaction among pioneers who sign up for the service. This too will pass as the providers gain footholds in their market penetration and service areas.

All in all, the current situation carries a catch-22: The carriers and providers need users to sign up for the service so that they can deliver more benefits and coverage. Yet users will be frustrated as the deployment takes place and will tend to migrate back to wired dial tone, since this is an older but proven technology. As the definition of personal communications is playing out, the world is wrestling with not only the service, but how to name it. As already mentioned, several acronyms are being bantered about to describe these offerings. The vendor community has begun to adopt two: personal communications services and personal communications networks (PCS/PCN). Variations of the service and the offerings have cropped up everywhere. Thus the dilemma.

Cordless Telephone Technology

The technology decisions in this arena are equally perplexing. As noted earlier, the typical frequency division services (analog transmission) in multiuser wireless communications do not allow for frequency reuse. Further, any attempt to use a narrower band for transmission would only place a limit on the services anticipated for use. Consequently, the

TABLE 5.1 Technologies Being Considered for PCS/PCN

Capability	Technology			
	TDMA	E-TDMA	CDMA	N-AMPS
Frequency reuse gains	3×	10–15×	20×	3–5×
Digital	Yes	Yes	Yes	No
Services to be offered	Voice Data Paging	Voice Data Paging E-mail	Voice Data Paging E-mail Radio determination	Voice Data Paging
Costs	Low–medium	High	High	Low
Quality	Fair	Good–excellent	Good–excellent	Good

industry has been wrestling with what technology to deploy for the future. Again, the costs will be astronomical when a nationwide or worldwide network service is installed. Therefore, time division multiple access (TDMA), enhanced time division multiple access (E-TDMA), and code division multiple access (CDMA) are all being considered by the various future providers. Each of the vendors or purveyors of the service could theoretically take a different approach.

The various technologies are summarized in Table 5.1. The table compares the respective frequency reuse gains, the digital versus analog capabilities, the services that are likely to be provided, and the potential costs and quality. One added comparison involves narrowband advanced mobile phone service (N-AMPS), which is being deployed on a trial basis by some of the cellular carriers. Since the variations will not be fully compatible, a change in the transmitted signal or in the format of the information will be required if an internetworking application is used.

Cost penalties exist with the various degrees of technology that are selected. The most expensive solution appears to be CDMA, which is still new and will require a total change of the equipment and techniques used.

The Concept of PCS/PCN

Visualize how this system might work! Wired telephone services from the local exchange carriers will still exist. Therefore, the wireless local loop in the dial tone area will still use a pair of twisted wires. However, the personal communicator will act as a cordless telephone within the business or residence. Figure 5.1 shows a cordless tele-

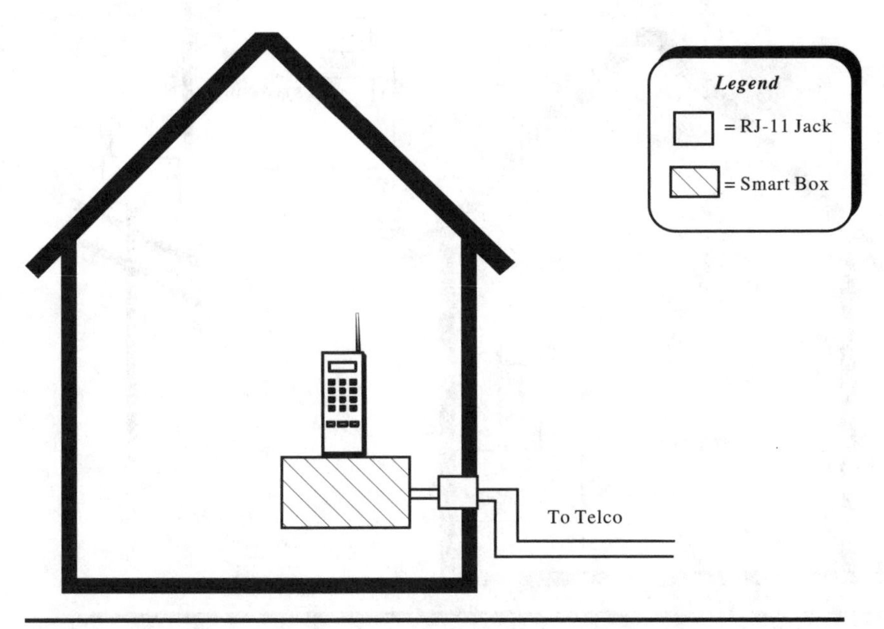

Figure 5.1 The personal communicator placed into a charger or smart box acts like a standard cordless phone when used with telco wires.

phone arrangement with a smart terminal attachment. This smart box will allow the user to place a personal communicator in the charging unit at the location. While in the charging unit, the set will act like any cordless phone. A call coming into the telephone will travel across the local telephone company wires and ring into the network-attached box. When the user picks up the phone and answers the call, the set will become cordless. Figure 5.2 shows the same scenario without a pair of telephone wires coming into the business or residence. In this scenario, the set is 100 percent wireless. With a smart interface in the location, a call to the business or residence will come in across the airwaves. The primary difference here is the absence of telephone wires.

Taking this one step further, when the telephone user leaves the primary location, the set will be taken from the charging unit and possibly clipped onto a belt loop or placed into a pocket, briefcase, or purse. As the user steps out into the open space of the neighborhood, the set will immediately begin communicating with a local cell. This unit, provided by the local telephone company or other supplier, will be a microcell possibly mounted atop a light pole or telephone pole (see Fig. 5.3). These cells will be wireless to and from the personal communicator, but the individual cells will be hard-wired along the pole line via fiber, coaxial cable, or twisted wire. The wired facilities

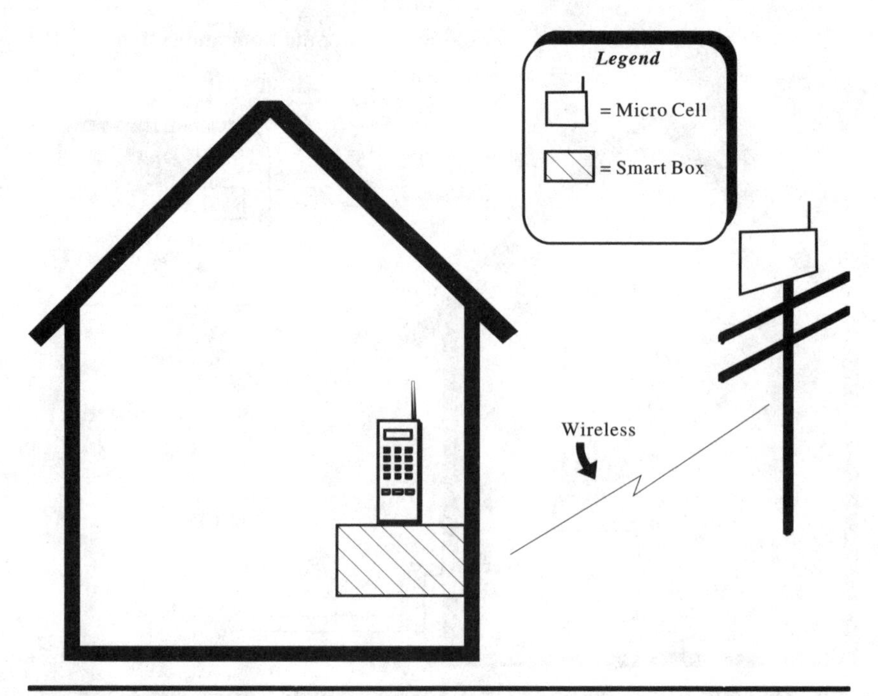

Figure 5.2 In the wireless arena, the communicator is connected to the telco via microcell technology.

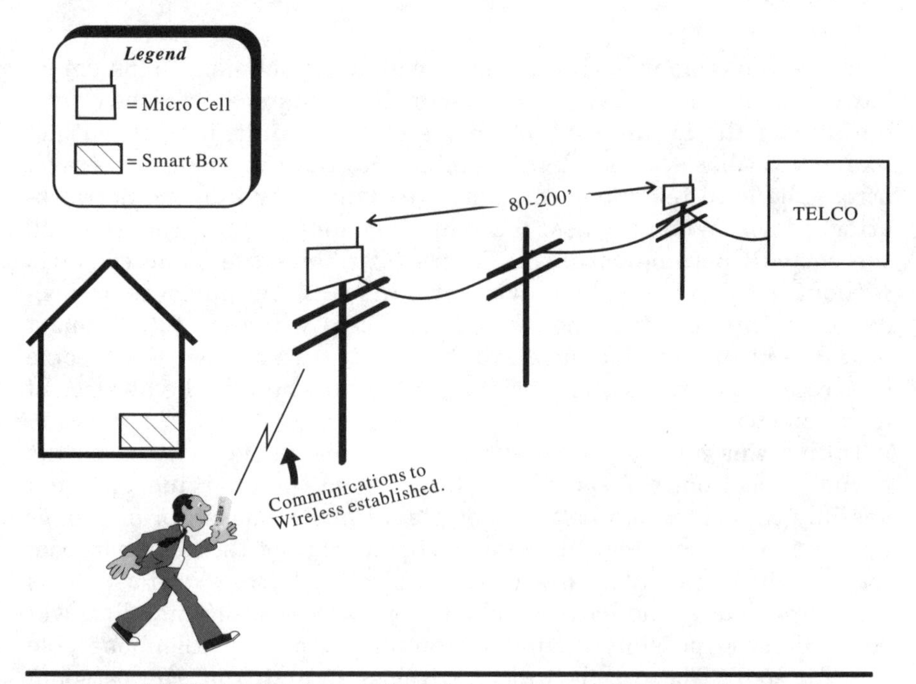

Figure 5.3 Once the set is removed from the smart box and the user leaves the locale, an instant communication is established via the microcell to alert the network that the user is mobile.

will be carried back to the telephone company's central office. In this particular scenario, the telephone company will have an edge over its competitors, since it already has the poles installed and owns the rights-of-way to the poles.

The user's set will continue to communicate with the local microcell as the user proceeds down a path or walkway. As the user gets close to a boundary of the cell, a handoff will be required from one cell to another. Each microcell will cover a limited distance to begin with, possibly 80–200 feet. This implies that handoffs will happen on a frequent basis—every 80–200 feet—as the user continues down the path. When the user enters an area where the poles are no longer around, the microcells will be mounted as telepoints along the way. A telepoint is a wireless access mounted to a building. In Fig. 5.4 the telepoint is mounted on the side of a building. In Fig. 5.5 the microcells are mounted on the tops of buildings with a radius of 200 feet between cells.

Each telepoint cell will be managed by a central control device, with a cell serving 20–25 simultaneous calls and a controller managing 32–64 cells. At some future point in time these controllers will likely manage up to 96 cells, each controlling 20–25 simultaneous calls. This amounts to a single control device managing 1900–2400

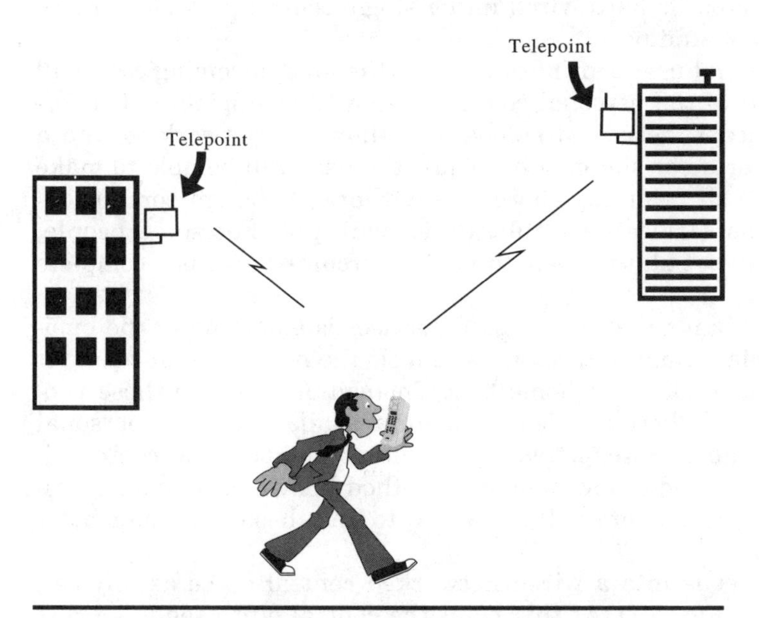

Figure 5.4 When no pole lines are available, telepoints can be mounted on the sides of buildings.

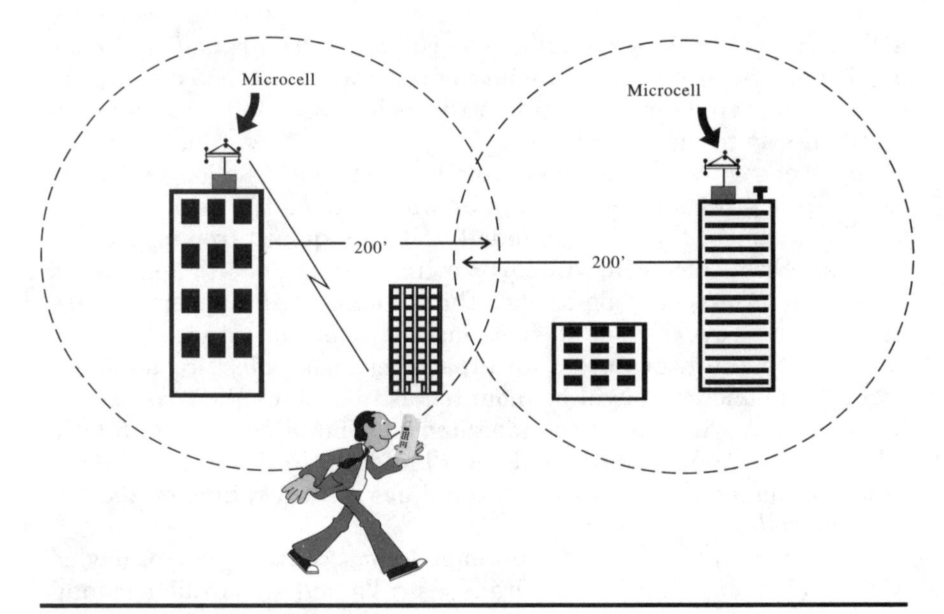

Figure 5.5 Microcells can also be mounted on the tops of buildings, with each cell serving a radius of 200 feet.

concurrent calls. Imagine the efficiency of this concept. Figure 5.6 shows 96 microcells hard-wired into a single control point. The numbers can be astounding.

As the original user continues through the area, incoming calls will be forwarded to the personal communicator. This single-number forwarding concept, described below, will allow the network to find a user wherever he or she may be. Thus, the user will be able to make or receive a call from anywhere to anywhere. Although some users may find this service desirable, others will not. For some people, being available at all times and under all circumstances is a complete annoyance.

The fact remains that a new breed of user is emerging in the business and social communities, one who feels the need to be in constant contact with the office or home base. Instead of rebuking these people, we would do better to leave constant availability as a personal choice. Through a newer networking concept, the personal communications service, and a new signaling method known as signaling system 7 (SS7), the network will know how to find these users no matter where they are.

A call proceeds into a wired networked central office as any call would do (see Fig. 5.7). At this point the central office sends an SS7 packet across the network to look up a database of how to handle

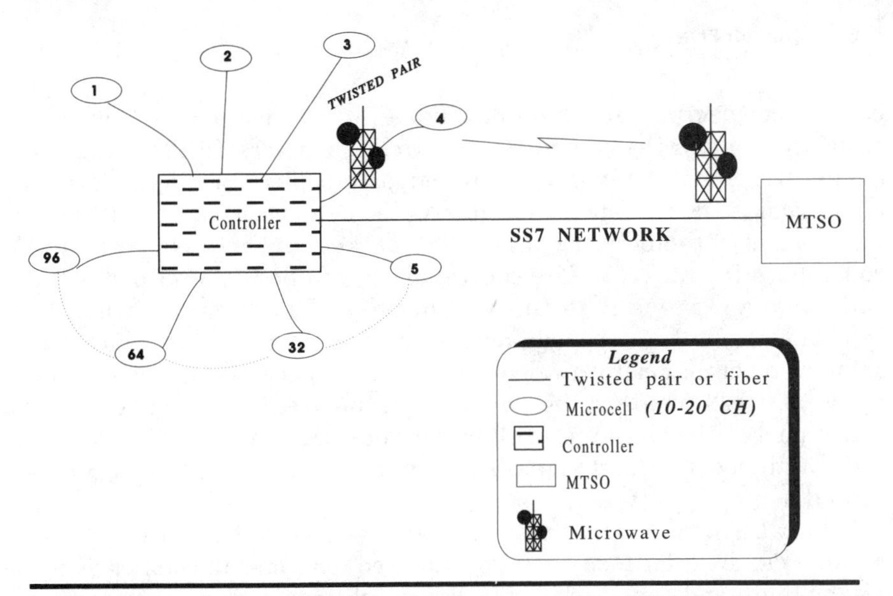

Figure 5.6 A single controller will be able to manage up to 96 microcells, with each cell handling 20–25 calls.

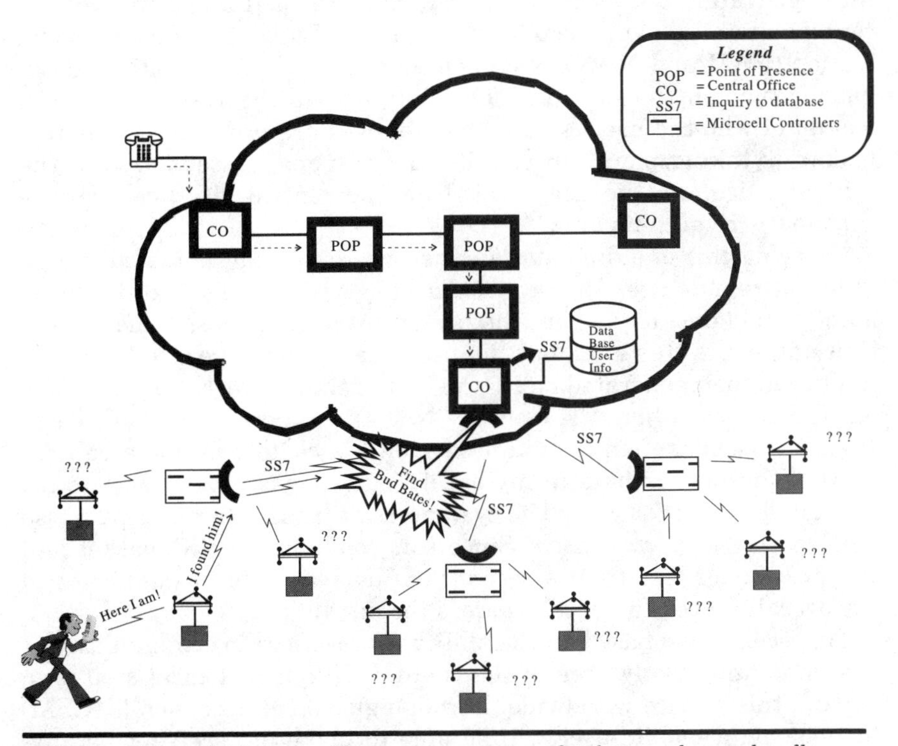

Figure 5.7 The local telco network issues a query to a database on how to handle an incoming call. The database responds by issuing a SS7 packet to poll local controllers to "find Bud." The response comes back, and the call is sent via wireless.

calls to a specific user. The database inquiry comes back with a response to hand the call off to the wireless carrier. (In this case, the same carrier may be involved. The important piece of information to be derived is the transport system.) As the call is sent out to the wireless portion of the network, a new SS7 packet is generated to all local controllers in the area. The controllers are polled to find out if the called party's phone is in the "on" mode, and if so where it is. The local microcells issue a page across the paging channels, much the same as a cellular network does. A paging response is then sent from the local user's phone to the controller in the area. Assuming that the called party is local, the call will be sent out immediately. The glory of a dynamic location and signaling system working together is represented here.

Let's take the scenario a step further. Assume that the user gets on a plane destined for a city several hundred (or thousand) miles away. As the flight ends and the user walks out of the aircraft, the user's set will immediately start to broadcast an alert message to the local wireless node closest by. In this case, when the remote controller receives the registration packet from the set, it will pass it along to the local central office. The local central office, not recognizing the user's registration (NAM and MAN), will send an inquiry to a centralized database server on the network. This inquiry will be in the form of a lookup of who the user is and what services are provided. While this lookup is taking place, an equally fast response is sent to both the central office and the user's set. Now the central office can service this new arrival in its area.

Carrying this one final step further, assume a call is dialed from a different remote user looking for our busy traveler. The call is generated to the home city where the person usually resides. A message is sent out across this area, but the user's set does not respond. An SS7 inquiry is then generated out to the rest of the network: "Does anyone out there know where the user is?" Now the remote central office can "hear" this packet inquiry and respond back to the home central office: "The user is here in my area! You can pass the call on to me and I'll deliver it for you." The system now can handle the traveler, *so long as the set is turned on.* Since this combination SS7 packet and response takes less than a second, the delay will be insignificant in the overall processing of calls across the network.

This scenario is based on the ability of two users to communicate at any time and in any area of the country. The next logical step is to expand this service worldwide. Through global services mobile (GSM) technology, a mobile user will be able to make or receive calls anywhere in the world. This expansion will take some time, since worldwide standards are being implemented differently and different technologies are being used around the world. As international standards bodies begin the final definition of how these systems will internetwork, the pioneers in the industry—providers that have already begun to deploy their services—will have to come up with a gateway service or else acquire new equipment. The changeover may also entail some dual-mode operations for a short period of time, with end users relying on existing equipment until they become willing to discard it in favor of the new standards-based equipment.

Function of Personal Communications

Serious efforts have always been made to eliminate the "telephone tag" problem within the business community. Instead of relying on a system of leaving messages and waiting for return calls, the industry has attempted to provide the connection directly. However, whether the user was in an office environment or at a mobile site, the limitations of the telephony world still prevailed. Too often, the user was relegated to a pair of physical wires. To overcome some of these problems and increase flexibility, cordless phones were introduced in both businesses and residences. Unfortunately, the same problems reigned—distance limitations. By reducing the amount of wiring needed in an office environment and introducing paging systems (loudspeakers within the office) and personal pagers (wireless notification services on a radio-based system), businesses felt that added flexibility and mobility could be achieved. Yet human nature crept into the picture, with individuals pleading that they were out of range and did not receive a page, or that the batteries to the portable pager or telephone set were dead. Hence, the problem was magnified. Frustration grew among management, customers, and clients.

The real intent of a telephone call is to connect two people who want to talk to each other, not to leave a paging or voice-mail message. Voice mail allows only one-way communication; leaving a message. Nothing has really been gained through this technology. Further, many organizations have learned that the voice-mail system, intended originally to assist in the passing of information, quickly eroded into an excuse never to have to answer the telephone again (not in real time, anyway). The individual could review the message and selectively decide whether to return a call. Pagers with alphanumeric displays allowed the caller to leave more information while trying to reach a party, but again this lent itself to screening, with the recipient evaluating the content of the message and when or whether to return the call.

Clearly, these attempts to put two people together were only preliminary steps toward the ultimate goal. The next logical step was to replace the paging and voice-mail systems ("telephone tag") with a device that could be used for real-time connectivity. As noted above, the cordless phone was the first attempt at such a replacement. Despite popular opinion, the cordless phone was in fact very successful, in a limited context—namely, the residential and small-business market. Overruling industry concerns about interference, lack of security (eavesdropping), and lack of privacy was the unabated desire to be untethered. Retail organizations, distribution and warehousing facilities, and small branch offices all found the cordless phone to be an improvement over the hard-wired environment. As the cordless movement swept the world, legal issues of eavesdropping, smuggling, and interference had to be dealt with. Single-line cordless telephony operated in the shared, unlicensed frequency ranges. Only a licensed (or controlled-device) technology could eliminate the problems of eavesdropping and interference. Hence, the personal communications concept evolved.

Early attempts at personal communications by various purveyors, under experimental licensing agreements from regulatory bodies around the world, met with mixed results and a confusing array of offerings. The situation called for another round of discussion about the packaging and delivery of the services. Personal communications is the attempt to rectify all the ills of past experiments and trials by delivering the call directly to the intended recipient. It is an oversimplification to think that a portable communicator will solve these ills. Personal means that the individual, not a company or department, will be reachable. Still, the decision remains with the individual whether to accept a call, forward the call to another location, allow the call to route to an answering machine or voice-mail system, or ignore the call entirely.

The fact that personal communications means that a recipient can accept or reject a call is only one part of the equation. The service also has to offer a much broader range of services and a much wider area of coverage. The personal communicator must be able to connect anywhere in a building, city, state, or country. Further, the service should include paging, E-mail, data transmission, facsimile traffic, and other types of connections that would normally be provided in the wired office environment. In short, the personal communications concept must include any form of telecommunications.

The business world is continually shrinking as users deal with customers and clients on a more global plane. Time-zone differences get in the way of conducting business with foreign or domestic customers. To overcome the problem, the user may have to stay late (or arrive very early) at the office for a call, fax, or other communication. This inconvenience leads to frustration, wasted time, and lack of freedom from the business. Also, today's new breed of superworkers (yesterday's workaholics) feel the need to be reachable at any time. Whether these workers are on vacation, traveling on business, or just at home relaxing, they don't ever want to miss that important contact. Consequently, the time constraints and demands on their social lives are beginning to require a new form of communications availability.

Doctors, lawyers, senior executives, and other professionals feel the need to be in contact with their offices around the clock. They place high value on being reachable at a moment's notice, particularly in light of the social, legal, and moral implications of the work they perform. These people cannot just turn off the clock at the end of the traditional business day. They must be accessible in order to handle on-the-fly decisions or assist in the personal crises of others. This new form of communications user wants and demands access to the network 24 hours a day.

Evolution of Personal Communications

Since its inception in the early 1980s, the wireless personal communications concept has mystified even the industry experts. What services could be provided at a reasonable cost, be ubiquitous enough to serve the masses, and still produce a profit? Further, while voice telephony is the primary goal of wireless services, newer applications are constantly surfacing. New systems should provide for the following user-selectable options:

- Voice dial-up services
- Data communications at various speeds, but minimally at 9.6 Kbps
- Paging and alphanumeric messaging on a pager
- E-mail access and file transfer capabilities
- Radio-determined vehicle location services
- Personal digital-assistant services for calendaring, mail, and contact management
- Facsimile at Group III speeds or higher
- Access to higher bandwidth applications, including multimedia
- Message center operations to leave or retrieve messages
- Telemetry for process control and alerting services
- Dial-up video in a slow scan mode (7–10 frames per second)

Each of these features is explored in the discussions below. It is important to note that the wireless communications network is somewhat band-limited. Reuse of the radio frequency spectrum is an added challenge. To provide these services, the wireless telephony world has had to undergo several migrations and evolutions.

The cordless telephone one (CT-1)

The cordless telephone one (CT-1) is the domestic telephone set that we have grown accustomed to over the years. A base station plugs into the home phone socket (the RJ-11 jack) and a power supply is connected to the user's standard electrical service. The CT-1 allows for a wireless connection between the cordless handset and the base station, which is plugged into the wall outlet. Intercommunications between the base and the handset are performed over radio waves, allowing the user to walk from room to room, or office to office. Originally, distances were limited to a couple of hundred feet from the base to the handset. Today distances of up to 1000 feet, and with amplifiers up to several miles, are available. CT-1 (also known in the international arena as CEPT-1) is a European standard for this type of set.

The primary problem with the first cordless phones was interference. The analog radio transmission associated with CT-1 sets could cause significant interference to radio equipment in the area, as well as to cable TV services in the immediate vicinity. Because few frequencies were available for use, frequency sharing was overdone. User frustration with interference, jamming, and loss of privacy and security (eavesdropping) led the industry to seek an alternative.

The cordless telephone two (CT-2)

The second in a series of evolutions was the cordless telephone two (CT-2), which originated in the United Kingdom. This service was based on the UK implementation of telepoints at which cordless pay phones, smart phones, phone points, and "rabbits" (product lines used to clone the phone points that are easy to install and inexpensive) could be introduced. The CT-2 was the first attempt to introduce a digital transmission system into the wireless phone business. There were some definite miscues with this technology, since it was introduced as a one-way transmission system. Calls could be made but not received, limiting the service to half of what the world wanted.

Eventually, CT-2 enhancements allowed for two-way calling services in office environments, but only at very limited distances from the telepoint. Another drawback was that the CT-2 could not make call handoffs from cell to cell, as do other cellular models. Once engaged in an active phone call, the user had to stay within a certain distance of the telepoint. Beyond that distance, the call would be cut off. It was as simple as that.

The cordless telephone two-plus (CT-plus)

Overcoming a problem is relatively easy in this arena, and the "plus" system allowed for the upgrade of CT-2 technology to accommodate handoffs and the ability to roam from one cell to another. In addition, the CT-plus offered two-way telepoint service, but at the expense of the system's overall capacity. The system throughput was literally cut in half or in quarter depending on the selections made.

The cordless telephone three (CT-3)

In 1990, L. M. Ericsson Co. introduced a new technique that moved away from the CT-2 system. The technique was aimed at providing a digital two-way service to the masses. CT-3 requires the installation of small cells, covering limited distances, in the area. These microcells and picocells are designed for the smooth and efficient handoff from one cell to another.

The seamless operation between cells is the cornerstone of CT-3 technology, which uses radio frequency services to provide cordless telephony in very crowded areas. In its design, the handoff of the call between the base station and the cells is totally transparent to the user. Feature and transport transparency is an important factor in gaining user acceptance of any technology. Like the CT-2 equipment, the newer CT-3 accommodates a telepoint application as well as wireless telephony in the business and residence. In both environments, the CT-3 set offers the advantages of cell-to-cell roaming, seamless handoff, and two-way calling.

Digital European cordless telephony or digital European cordless telecommunications (DECT)

As the industry wrestled to introduce a cordless telephony world, the European standards bodies were busy at work developing guidelines for all to follow. In early 1992, the DECT standard emerged and was accepted by the member countries of the Council of European PTTs (CEPT). DECT is based on the same concept and technology as CT-3, but uses a different set of radio frequencies plus other, minor variances. Functionally, the two systems are identical.

The Range of Technological Offerings

With all the emerging technologies, industry vendors and standards bodies still have to address the problem of frequency reuse. The resources of the radio frequency spectrum are finite. No one solution will obviate the need for prudent spectrum management. As they battle to provide services to the masses, the newer techniques will also have to face the issue of where the spectrum is going to come from.

In the United States, the Federal Communications Commission has attempted to free up spectrum from the limited bandwidths available in order to meet the needs of and demands for various service options. No one solution appears to be effective enough to reclaim the total spectrum for the dynamics of allocation. The techniques being worked on globally only add confusion to the industry. Recall that there are several ways that the spectrum can be used, ranging from simple frequency division to time division and now code division. As the new standards are developed, the engineering resources within the industry must evaluate them carefully to overcome the problems posed. If several different techniques are implemented, incompatibilities are bound to arise.

As noted above, the CT-2, CT-3, and DECT services are functionally similar. However, they use a different technique to modulate the information onto the radio frequency spectrum. Thus, the more critical concern is the way the equipment will gain and control the speech channels through a dynamic method of channel allocation. CT-2 is designed around frequency division multiple access with time division duplexing (FDMA/TDD). The DECT and CT-3 services are designed around time division multiple access with time division duplexing (TDMA/TDD), although slight variations exist in their implementations.

Frequency division multiple access with time division duplexing (FDMA/TDD)

FDMA splits the full bandwidth into separate channels within the total frequency spectrum being allocated and used. Therefore, the capacity of the system is a function of the total bandwidth available. If more capacity is needed, more channels must be allocated—hence, more spectrum and bandwidth. The rate of consumption in a frequency division system is much higher than in other systems. FDMA is a proven technology and has some merit; it is compatible with established spectrum-use techniques. However, the equipment needs of FDMA can make the portable systems more costly, and the technique can be complex to administer.

The CT-2 systems in the United Kingdom were allocated 4 MHz of bandwidth, subsequently divided into 40 channels of 100 KHz each. This was the functional equivalent of 40 simultaneous voice channels. These channels used a time division duplexing (TDD) scheme. To ensure that no interference would occur, the capacity planning for these systems had to take into account the channels in use within adjacent cells. Otherwise, two calls on the same channel could begin to interfere with each other.

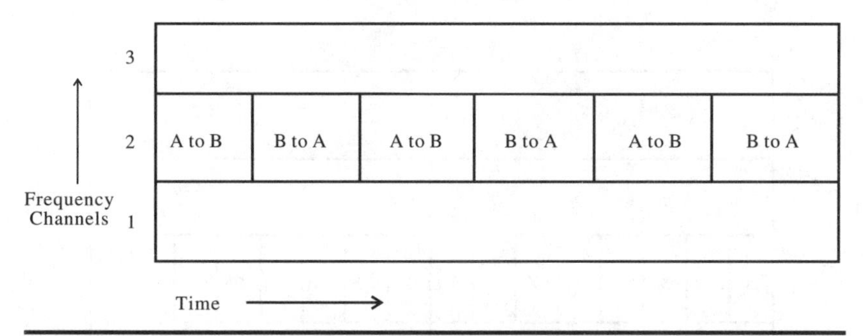

Figure 5.8 FDMA systems split the bandwidth into fixed channels. A call occupies the complete channel and splits it into time slots for transmit or receive (TDD).

As with the cellular systems discussed in Chap. 4, the cell-to-cell handoff is a key concern in an FDMA call. During the initial call setup, the equipment scans the available channels. Once a free channel is found, the set is locked onto the channel on a specific frequency. Figure 5.8 shows how the bandwidth is split between time and frequency. Although this technique has been well established over the years, the handoff from cell to cell must be carefully coordinated. Otherwise, the call is likely to be dropped as the user moves from cell to cell, since a free channel is not available. When the density of users increases, channel coordination can become a serious problem. The FDMA system uses a dynamic channel allocation to gain access to the system. Once this occurs, the terminal device has an established conversation path and occupies the entire frequency allocated to this channel. Therefore, the terminal cannot monitor other channels in its own cell or in adjacent cells. It is an island unto itself for the duration of the call and stands alone in the coordination of handoff from cell to cell.

Time division multiple access with time division duplexing (TDMA/TDD)

TDMA allows access to the full bandwidth of the frequency spectrum, divides it into small time slots, and then allocates speech to these time slots. The typical bandwidth allocated to DECT is 20 MHz (1880–1900 MHz); 10 carriers are established. Each carrier is divided into (12 $\times$ 2) 24 time slots, thus providing 120 access channels. CT-3 equipment is 4 MHz. The bandwidth is then subdivided into 4 separate 1 MHz carriers. Each 1 MHz carriers is divided into 16 (8 $\times$ 2) time slots. The time slots are paired providing 32 access channels. One slot will carry conversations from the base station to the terminal device, whereas another slot will carry conversation from the ter-

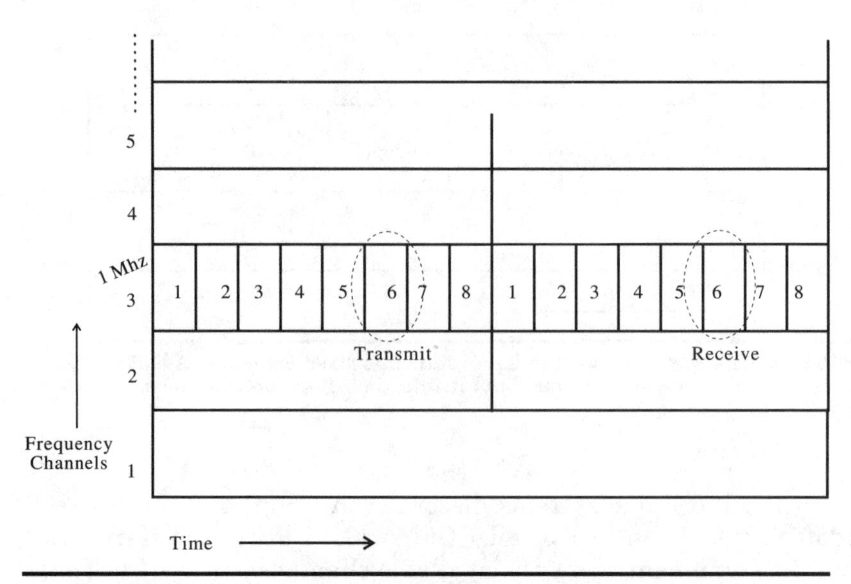

Figure 5.9 TDMA systems split the bandwidth into channels and then split each channel into time slots. The call uses only two time slots, alternating between transmit and receive (TDD).

minal to the base station. Figure 5.9 is a graphical representation of the time slot formulation.

The CT-3 16 time slots create a full duplex capacity for 8 channels and are formulated into a frame of information. Each time slot is only 1 millisecond (ms) long, and the frame is 16 ms long. In Fig. 5.10, the time frame is shown as a full duplex operation. Therefore, a fully loaded system will allow access for 32 full duplex channels per cell (8

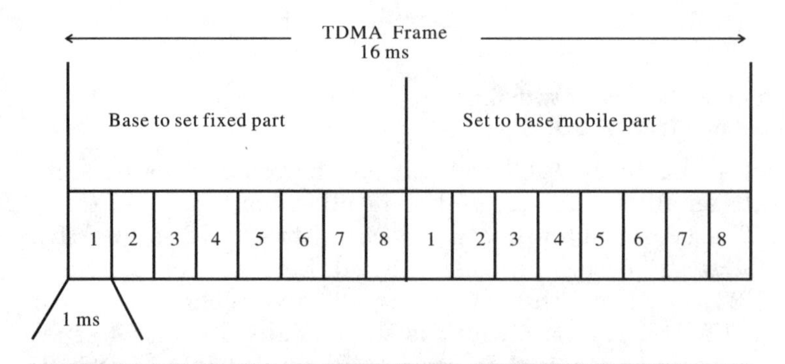

Figure 5.10 The TDMA time frame is 16 ms. Each time frame is broken down into 16 time slots: 8 for transmit and 8 for receive. Each time slot is 1 ms long and carries 32 Kbps.

duplex channels per 1 MHz segment times 4 MHz). Since the TDMA terminal will use only a portion of the time allocated to it during the conversational mode, it will be able to monitor other channels within its own cell as well as those in adjacent cells on all the frequencies assigned. The information received in these off times can be stored in the terminal memory for later access and retrieval, or it can be used "on the fly" to access new services real time.

As the user moves around, the coordination of the handoff from one cell to another can be seamless, since the other channels can be monitored simultaneously. Since this handoff will take place within one time frame, or 16 ms, the user will never know the difference. This feature of microcells and picocells is important, since multiple handoffs may be necessary in a relatively short period of time. A further benefit of the TDMA service is that third-party information can be delivered to the terminal during a conversation. This means that while the user is talking, another party can leave a callback message for the user. An E-mail or a paging message can be delivered for later access while a conversation is in mode.

The TDMA channel is basically set up as a digital-adaptive differential (delta) pulse code modulation (ADPCM), or 32 Kbps of digital bandwidth. Since the channel is duplex, 32 Kbps are allocated for transmit and 32 Kbps for receive. However, if dynamic allocation of the bandwidth is needed, time slots can be concatenated with a yield of 64, 128, 192, and 256 Kbps for data transmission or any other service that emerges (see Fig. 5.11). The transmission could theoretically be made simultaneously as a voice and data call from the single set. Imagine the power of such personal communications with this form of digital transmission available at the personal communicator.

As already mentioned, the DECT and the CT-3 standards differ slightly. Table 5.2 compares the two techniques. Some of the primary differences are the length of the time slot, the channel bandwidth allo-

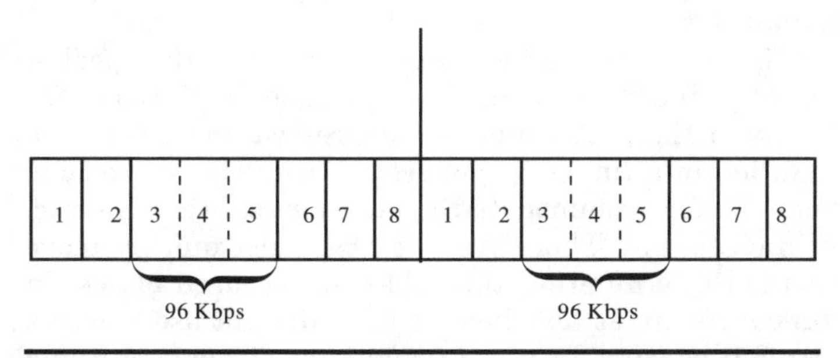

Figure 5.11 By superrate muxing, 128, 192, and 256 Kbps data streams can be achieved.

TABLE 5.2 Differences Between CT-3 and DECT Standards

Description	CT-3	DECT
Length of frame	16 ms	10 ms
Carriers	4	10
Duplex channels per frame	8	12
Access channels	32	120
Time slots per frame	16	24
Yield data rate	640 Kbps	1.152 Mbps
Bandwidth allocated	4 MHz	20 MHz
Radio frequency used per carrier	1 MHz	1.728 MHz
Digital data rate	32 Kbps ADPCM	32 Kbps ADPCM

cation, and the overall data rates available. The figures given are for illustrative purposes only, since rates and services change all the time.

Enhanced time division multiple access with time division duplexing (E-TDMA/TDD)

Another technique being considered for deployment in the PCS/PCN world is enhanced time division multiple access with time division duplexing. E-TDMA seeks to improve on the TDMA techniques currently in use by providing a more robust allocation of the bandwidth with access for more users. As such, E-TDMA needs more work before it becomes an accepted technology in the public marketplace. Expectations are that a tenfold increase in the number of simultaneous users on a channel can be achieved. In E-TDMA, a single channel will allow multiple users to share a single frequency through a digital time slot interchange (DSI). Recall that most voice applications are one-way alternating conversations, in which one end of the transmission path is in the transmit mode while the other end is in the receive mode. Therefore, half the path is idle, at least for a good amount of the time. In fact, during the transmit mode, only 10 to 20 percent of the circuit is actually being used.

Human speech is characterized by frequent pauses as the speaker breathes, thinks, or ponders a response to a question. With E-TDMA, all this dead space on the path could be compressed and used for a different conversation in a time division sequencing. The sequencing must be quick enough to accommodate the need for real-time interactive transfer of information. Therefore, the DSI must quickly move pieces of information into different time slots as the need arises. In Fig. 5.12, the time slots are established for five different users on the circuit. There are 8 time slots for the transmit side of the circuit and 8 time slots for the receive side of the circuit. The time slots are allo-

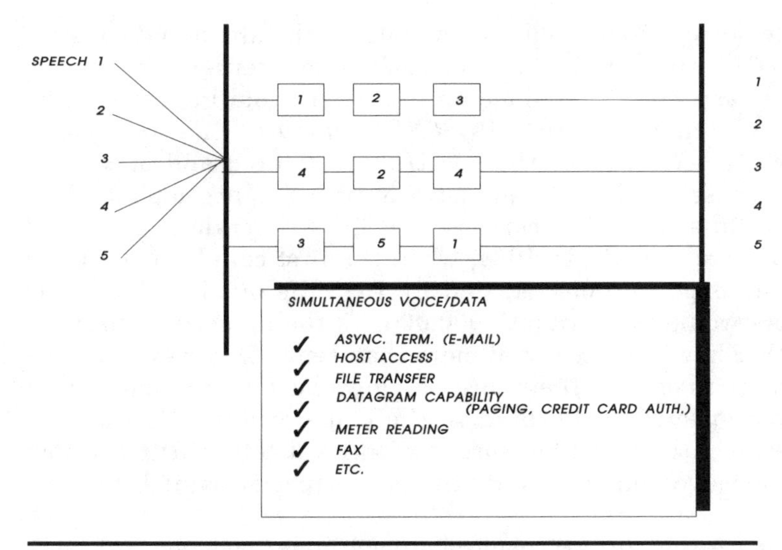

Figure 5.12 Using E-TDMA with digital time slot interchange. The user sample is placed in different slots in various frames.

cated for use by the transmitter in a sequence as needed. The arrangement takes care of the dynamics of the call, allowing the system to provide full duplex operation, or what appears to be full duplex operation, for the five users.

In Fig. 5.12, note that user 1's sample of the transmission is slotted in the first frame in channel slot 1, whereas in the second frame the same user is allocated into channel slot 4. The system will deliver this time slot interchange as needed for continuous transmission. As noted, E-TDMA is a new technique that will require further testing for clarity of transmission, responsiveness to circuit and user needs, and ability to control a cell-to-cell handoff when the time slot sequence is a moving target. Supporters are very strong on the ability of E-TDMA to serve as a primary tool for PCS/PCN in the future.

Code division multiple access with time division duplexing (CDMA/TDD)

One of the most talked-about technologies in recent years is code division multiple access with time division duplexing. CDMA was introduced to the industry by Qualcomm, a satellite carrier that had experimented with the technique in such applications as specialized mobile radio (SMR) services for vehicle-tracking systems. By consensus, the industry had already locked in on the development of TDMA systems and techniques. However, while the industry was figuring out how to deploy TDMA, Qualcomm entered the arena and suggested that CDMA would be far superior to all other alternatives.

CDMA is expected to produce up to a twentyfold increase in the effective spectrum for future PCS/PCN services. This increase is obviously attractive, since the spectrum is limited and maximum frequency reuse is what the industry is trying to accomplish. Just as in the cellular industry, carriers of PCS/PCN services are poised for phenomenal growth. In 1984, when the first cellular operations were begun, experts projected that by 1995 the number of cellular subscribers would approximate 900,000. To the industry's surprise, that number has been exceeded more than tenfold. However, as the cellular carriers bask in their success, they also face a problem with the limited spectrum allocated to their operations. Crowding in major metropolitan areas has forced significant cell splitting, which is very expensive and still limits the total capacity available to the carriers.

As the move to digital transmission gets under way, several techniques are being considered. As already mentioned, the options include the existing AMPS, the newer TDMA, E-TDMA, and FDMA, and finally CDMA. Another option in the mix, narrow-band AMPS (N-AMPS), seeks to introduce the concept of splitting at 30 KHz cellular channel into three 10 KHz channels. However, this is only a stopgap measure. The potential twentyfold increase in spectrum use offered by CDMA is exciting to potential carriers. CDMA offers several other advantages over existing technologies:

- Improvements in the handoff process
- Improved security
- Enhanced privacy
- Higher voice quality

Clearly, these are advantages that are also desired in the PCS/PCN world. Table 5.3 summarizes the AMPS, TDMA, and CDMA systems

TABLE 5.3 Improvements of TDMA and CDMA over AMPS

Description	AMPS	TDMA	CDMA
Cellular bandwidth	12.5 MHz	12.5 MHz	12.5 MHz
Radio channel bandwidth	30 KHz	30 KHz	1.25 MHz
Calls per channel	1	3	36–38
Calls per cell sector	19	57	360–380
Capacity improvements over AMPS	1 ×	3 ×	20 ×

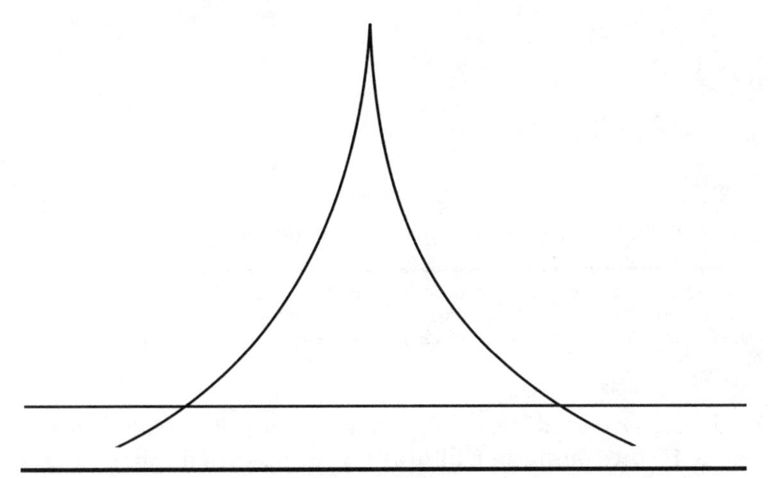

Figure 5.13 In conventional radio, the signal is concentrated at full power on a specific frequency.

and shows the density and capacity gains expected from the use of CDMA. From the number of voice channels available through cellular suppliers, a derived quantity of increased service from PCS can be interpolated.

Figure 5.13 shows how, in conventional radio, the signal is concentrated at full power on a single frequency so that the entire channel is occupied by a single voice conversation. The rise and fall time of the energy as it is modulated onto the carrier frequency uses the entire channel. Figure 5.14 shows how, in the CDMA spread spectrum, the energy of the conversation is spread across multiple frequencies over a rapid period of time. This frequency-hopping sequence is immune to the interference and noise that characterize frequency division channels. CDMA frequency hopping yields much greater reuse of the radio spectrum than the other techniques discussed. The direct sequence coding of the digital signal allows multiple conversations or data streams on the same frequency, since each conversation has a unique coded sequence.

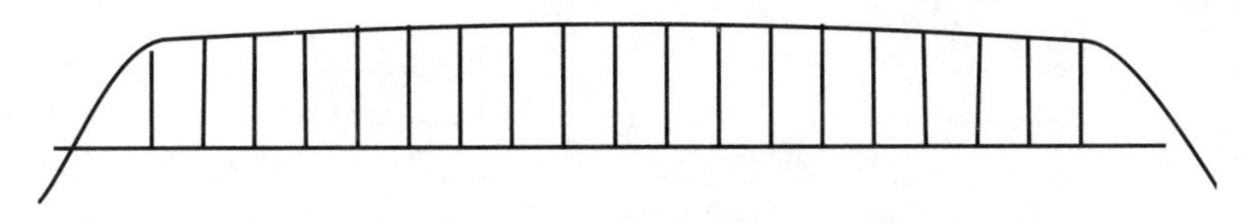

Figure 5.14 In spread-spectrum CDMA, frequency hopping is employed, spreading the signal across multiple frequencies.

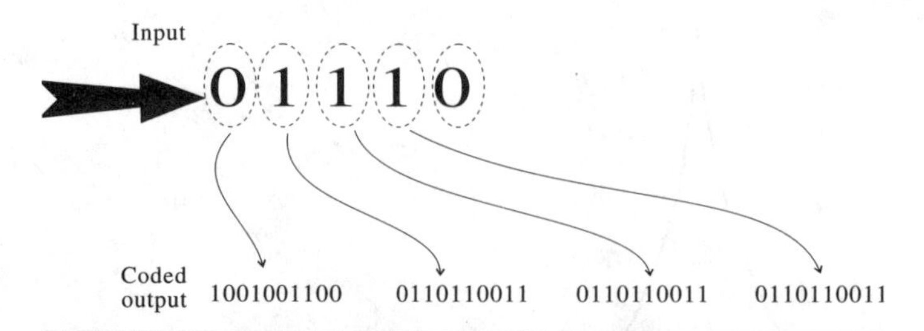

Figure 5.15 In CDMA, a direct sequence code is digitally applied to the signal. Multiple signals can be transmitted on the same channel simultaneously, since each has a unique coded sequence.

Figure 5.15 illustrates direct sequence coding. This particular sequence uses a 10-bit coding per digital signal as it is applied to the channel. The military has used coded sequences for years to prevent jamming of signals transmitted over the radio frequency spectrum. However, the military sequences are 100-bit patterns, as opposed to the 10-bit patterns used in CDMA. In Fig. 5.16, the sequence generator is introduced at the radio frequency (RF) modulator. In this case, the generator at the receiver end must be closely synchronized so that the generator at the receiving end can decode the sequence.

The PCS and Cellular Markets

As this discussion flows along, readers may well begin to wonder what the differences will ultimately be between PCS/PCN networks and the existing cellular networks discussed in Chap. 4. The answer is that there probably will not be a lot of technological differences between the two networks; the functionality and the population

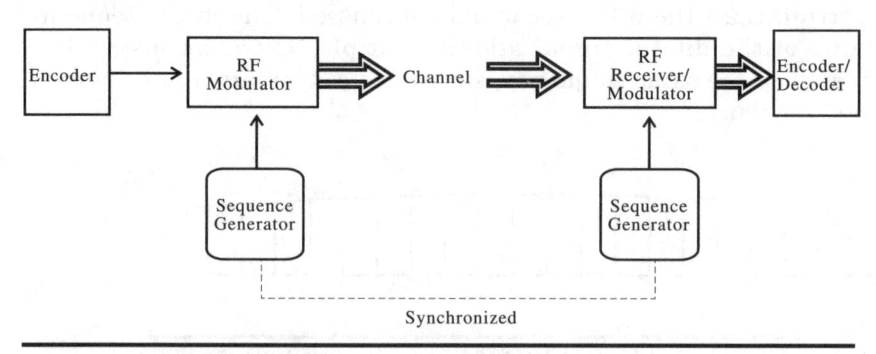

Figure 5.16 In spread-spectrum CDMA, security can be enhanced by using a noise sequence generator along with the actual input. The two generators must be synchronized so that the receiver can decode the signal.

served will likely be the differentiating factors. PCS/PCN has been touted as the replacement for the local telephone loop, or hard-wired service. This is not to suggest that all local wired facilities will disappear or be abandoned. Nothing like that will take place for at least a generation. However, as a secondary means of providing the single-number telephone concept, enabling the user to be totally mobile yet still be reached, PCS/PCN will fill the niche. On a higher-quality 64 Kbps standard, providing premium sustained service for the fast-moving vehicular-based user, cellular will offer the appropriate connectivity. Therefore, the two technologies and services should be viewed as complementary rather than competitive.

The cellular network will continue to support the office-in-the-car concept for voice, data, video, fax, and image transmission. The costs will continue to increase as this service becomes more robust and dynamic. For most casual users, or mobile users who need to be near a residence or office, PCS/PCN will suffice. Therefore, two separate target markets will still be evident. This is not to say that the providers of the service will necessarily be different. The cellular provider of today may well be the cellular and PCS/PCN provider of tomorrow. On an international plane, this has always been true. Local post telephone and telegraph (PTT) organizations were the only authorized providers. However, with the shift to competition in the local-loop bypass plus the global service offerings, several countries are beginning to make tenders to other carriers and suppliers. What was once the sacred monopoly of governments around the world is now breaking down to a competitive service offered by many private carriers. The various technological offerings have been presented in Table 5.1. Table 5.4 summarizes how the market may play out for the delivery of these two complementary products, and the number of players that may well emerge to provide this service.

Services and suppliers

In the beginning stages of PCS deployment, the carriers will be offering advanced services at lower prices. At this point the market will likely be two major cellular players and up to five PCS suppliers. As time goes by and newer technologies and services are deployed, the number of players in this market will be wide open. Consider that in October 1993, the FCC announced its intentions to provide PCS services through a series of auctioned licenses to approximately 2500 licensees. There are likely to be 11 national PCS providers, followed by 49 regionals. The rest will be localized in a specific metropolitan or rural area. The market penetration at this juncture will be wide open.

TABLE 5.4 Features and Suppliers of PCS/PCN in the Future

Item	Sustained premium	Digital high end	Microcell PCS/PCN	Mass-market PCS/PCN
Technology	Analog 800 MHz (AMPS)	Analog/digital Macro/microcell	Digital cellular PCS 1.7–2.3 GHz Microcell Wireless PBX	Wireless access to network Wireless PBX SS7/IS-41
Customer equipment	Vehicle-based Transportable Handheld	Vehicle-based Handheld Wireless PC	Shirt pocket communicator	Credit card phone Wristwatch Smart card
Service offerings	Analog voice Paging	Analog voice Digital voice Advanced messaging	Analog voice Digital voice Advanced messaging	Personal number ANI SS7
Cost/price	Lower than telephone service (⅔) High	Digital less than analog High	High initially Lower than cellular	Low Cheap
Market	Business Upper income	Business Professional Upper income	Broader range of users Residential/middle income	Mass market International
Driving forces	High revenues Reduced churn	Decreased costs More competitive pricing Capacities	PCN vs. cellular Existing operations Unsure	Market-generated
Number of suppliers	2	2	5	Wide open

SOURCE: *Cellular Magazine*.

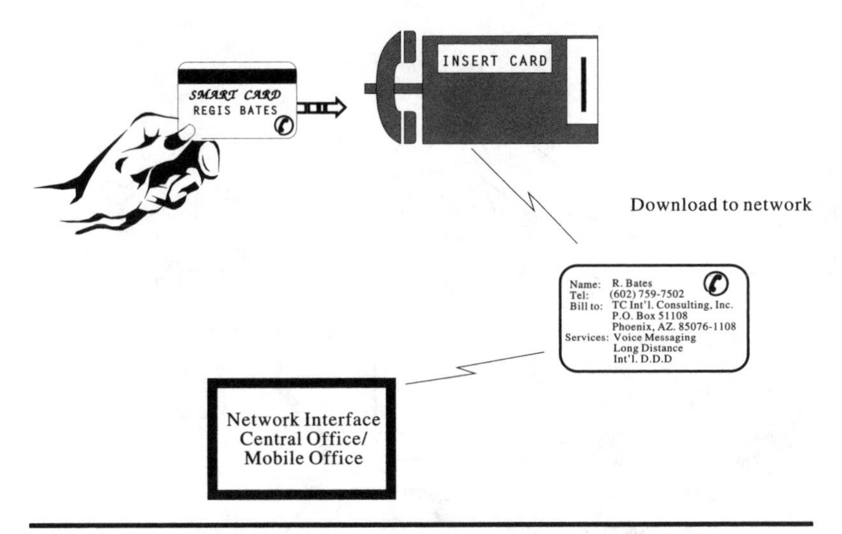

Figure 5.17 A smart card will be a newer evolution in the PCS/PCN world.

Further, the smart phone card (see Fig. 5.17) will emerge as the wave of the future to allow total connectivity anywhere and any time. This is how the telephony world will work in 1998 and beyond.

Several experimental programs are being conducted at this time and will likely continue through the end of 1995. Further, the first of the national licensees (a company called M-tel from Jackson, Mississippi) expects to have a nationally deployed system up and running by the end of 1994. The M-tel concept (see Fig. 5.18) uses a series of interconnected microcells across a larger backbone network to provide voice, data, text, E-mail, and paging services across the United States. A personal digital assistant—a handheld communicator—will be the primary access device into the M-tel network. Although this is not a fixed function, industry users are looking for a fully functional digital instrument that combines access to all the technologies mentioned. It will be amazing to see how the whole system rolls out.

Licensing

As noted, the FCC has signaled its intention to auction off licenses by mid-1994, so that the construction of the networks can begin. This is a departure from the lottery system used in the allocation of frequencies and licenses for the cellular industry. The government expects to raise $10–$12 billion through auctioning the spectrum. It seems that the airwaves are no longer going to be free, and that the government has discovered a brand-new way of raising capital without taxation!

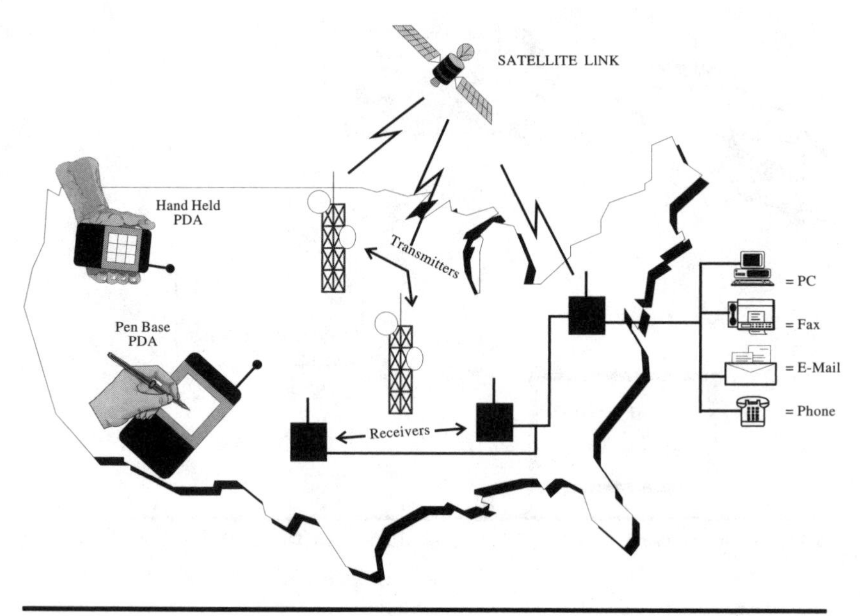

Figure 5.18 The M-tel concept for national PCS/PCN services is expected to be available in 1994. (*Courtesy of M-tel, Jackson, Mississippi, September 1993.*)

The FCC has obviously learned that the capital investments are going to be significant and that the strongest players will be the survivors. Therefore, in order to be granted licenses, bidders must show the financial stability and backing required to sustain large capital investments while return revenue streams are low. Each bidder must construct a decision tree to determine how best to deploy PCS/PCN. Figure 5.19 lists some of the questions that may require answers before such a deployment is undertaken. These issues begin only the decision-making portion of the process.

Time frames for deployment

As FCC licensing of PCS/PCN suppliers gets under way, the rollout of technology will become a key concern. Just when can it be expected, and how will the rollout occur? Table 5.5 suggests a possible sequencing, though only a visionary could provide a flawless timetable. The carriers and local exchange providers are all eager to supply the service, but timing remains an enemy. If a system is introduced too soon, the carriers will have to gamble on the technological decision; if it is introduced too late, the niche could already be filled. It is highly likely that the first player in an area will gain the market niche, with each

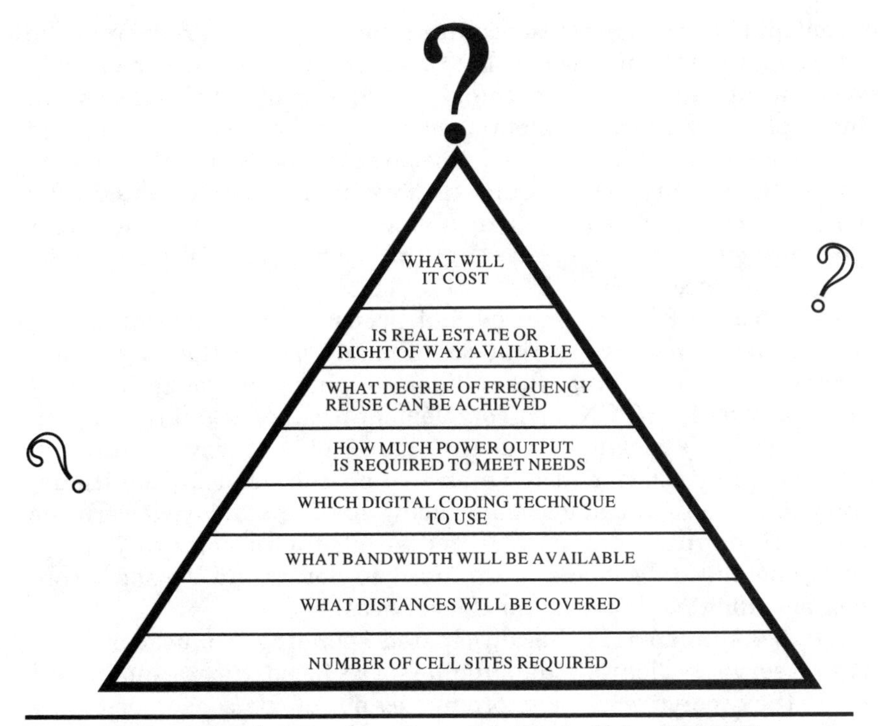

Figure 5.19 A series of decisions must be made before deploying PCS/PCN.

TABLE 5.5 PCS and Cellular Services for Near-Term Deployment

Microcell technology	Cellular operations
Coming within 2–6 years	Here now and being updated
Private network suppliers	Public network suppliers with near-ubiquitous access
Smaller footprint for cells; less expensive radio systems	Smaller footprint and cell sizes deployed through cell splitting
Costs high in the beginning, but will drop in future	Costs higher for premium services, but will drop as competition and PCS are introduced
Unlimited number of users	Limited number of users (possibly 14 million by the mid-1990s) but will improve as technology gains acceptance
Smaller set with more access ports, possibly the size of a credit card	Larger set with external antenna, but size is continually shrinking
Ubiquitous penetration by the turn of century	Limited areas of coverage, but higher-powered sets allow for more depth and penetration

new entrant merely trying to acquire a piece of the market from the existing player(s). This means that, again, return on investment will be somewhat diminished for the short term while the systems are being deployed. As already mentioned, if a carrier offers a service and the network is not fully deployed, user frustration will build, causing a higher-than-desired churn rate. If the system is implemented fully before service is offered, the provider will have to carry the paper losses until such time as users adopt the service. This is the catch-22 arrangement once again.

Although PCS/PCN technology and licensing are the focal issues, financial decisions remain that could easily get in the way of full deployment. For one thing, the cellular suppliers are going to be bidders to provide PCS/PCN as a complementary, and possibly competitive, service. This means that as they gain rights-of-way, acquire cell sites, and so on, they will have an eye on the need for additional equipment. The financial decisions facing suppliers today concern the costs of the equipment as well as which technology to deploy. Underlying these decisions is the need to determine if usage (consumer spending) will pick up for the services.

Table 5.6 summarizes the anticipated spending per user on both types of service. Clearly, the suppliers are faced with a quandary: How do they coerce or cajole more money out of the consumer to pay for their investments? This issue is particularly vexing for PCS/PCN providers, since statistics indicate that many users rent access to cellular networks, but use the portable or vehicular phone only for emergencies. Given a cost per installed cell and channel per user of approximately $1200, the carriers cannot afford to have users on the system who are generating limited to no revenue.

TABLE 5.6 Current and Future Spending by Service

Anticipated monthly consumer spending	
Typical telephone bill	$30
Anticipated PCS/PCN bill	$50
Typical cellular bill	$80–$220

Chapter

6

Comparing PCS and Cellular Technologies

The past two chapters examined the individual technologies to be used with both PCS and cellular systems. This chapter explores the subtle differences that will likely emerge. As the technologies are introduced to support the PCS world, they must display characteristics that distinguish them from cellular operations. If no differences exist, there would be no incentive to introduce the services.

What differences could separate the new personal communications services from the existing high-end cellular services? Clearly, one of the primary service differences will be cost. As already mentioned, although PCS will start out as an expensive implementation, the costs will have to fall dramatically or the product differentiation will not support the service. However, just how fast these prices will drop remains a mystery. The costs of cellular service are continuing to climb, at a time when the prices should be falling. The reasons are fraud—the invasion of the cellular network by unauthorized users—and, to a degree, the failure of authorized users to pay. Further, the costs of technology are falling, although at slower rates than originally anticipated. Whereas the first cell sites were installed in the $850,000 range, newer sites are being installed in the $650,000–$700,000 range.

A Seller's Market

The cost of real estate to locate cells has held somewhat stable, causing new cellular owners and operators to scramble for the appropriate rights-of-way and building space. The result is a seller's market in which cellular operators scurry around to lock in cell sites before the personal communications suppliers gobble up the remaining desirable space. The cellular suppliers have already had problems meeting their

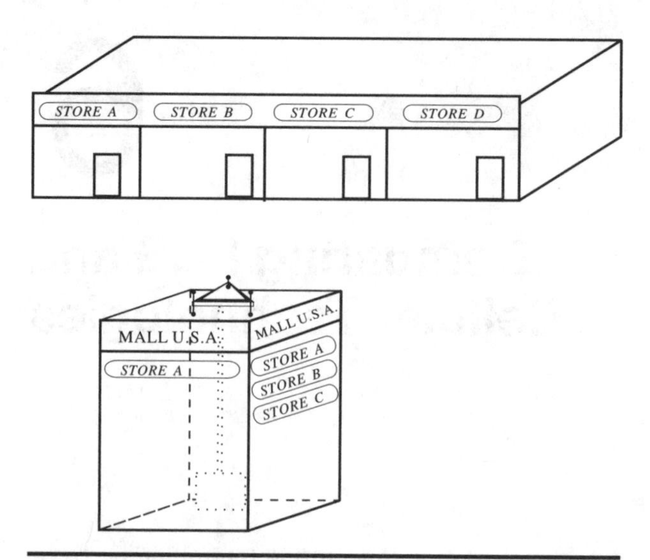

Figure 6.1 Cells are hidden for aesthetic reasons.

own needs and have had to modify the look of their newer cell sites, sometimes even hiding their equipment (as in a marquis in a mall) for aesthetic reasons. Figure 6.1 shows how some cellular suppliers have become creative in finding space and concealing the identity of the site.

PCS suppliers will be using much smaller site environments with their microcells; however, they may be faced with the same need to disguise their equipment. This can be done in the form of a microcell mounted on a rooftop (see Fig. 6.2) or a flat telepoint mounted on the

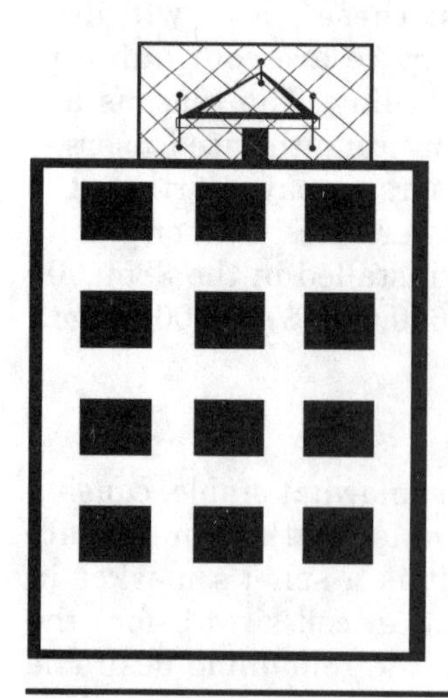

Figure 6.2 A microcell can be hidden behind a baffle on top of a building.

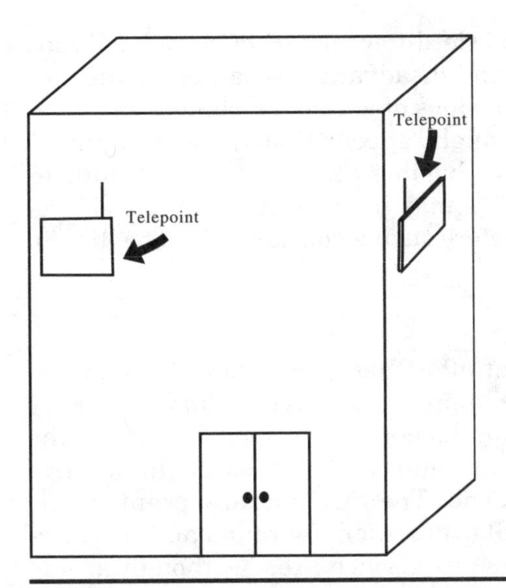

Figure 6.3 Aesthetically pleasing flat telepoints can be surface-mounted on the sides of buildings.

side of a building (see Fig. 6.3). Regardless of how equipment is housed, the needs of both suppliers will be the same. The PCS suppliers will have the disadvantage of having more cells to hide, since each microcell has a range of only a few hundred feet, whereas cellular operators enjoy a 3–5 mile radius. Table 6.1 compares these two technologies and services. The list is not exhaustive, but is an attempt to

TABLE 6.1 Comparison of Cellular and PCS Technologies

Cellular	PCS/PCN
High antenna and more space for site required	Smaller footprint for the microcell
Fewer sites needed to provide coverage	Many more sites needed to provide coverage (possibly a 20:1 ratio)
More expensive equipment ($600,000–$800,000 per site)	Less expensive cells ($20,000–$60,000 each) but many more needed
Higher costs for airtime (approximating $.35–$.65 per minute)	High costs initially, with airtime costs dropping rapidly by 1998 (down to $.10–$.20 per minute)
Higher-power output (3–15 watts), requiring more in-depth studies and ordinance approvals	Lower-power output (.1–.5 Watt), facilitating ordinance approvals
No major advantage in each city of operation	Possible right-of-way advantages because of telephone pole lines and existing rights-of-way

build on the similarities and subtle difference of PCS and cellular. From this table, added benefits and disadvantages can be extrapolated to show other concerns on both sides of wireless technology.

From the above discussion, it might appear that there is no incentive to be in the high-end cellular business and that the advantages will all go to the PCS carriers. This may be true, but other considerations will have to be addressed before such a conclusion can be drawn.

Frequency Bands

A considerable amount of diligent effort has been afforded the industries on both sides to allocate the appropriate spectrum in a frequency band. Further, use of the spectrum has been protected for each of the carriers so that interference is minimized. The task of finding frequencies is always a formidable one. The FCC initially granted permission for two competitive cellular operators (wireline and nonwireline) in each metropolitan and rural service area. Although at one time the FCC had allocated enough frequencies for a third competitor in the cellular business, the third set of frequencies was never issued.

The frequency band for the cellular operators is in the 800–900 MHz range, which is already becoming quite congested. Using this same band for the PCS operators poses additional difficulties. Therefore, the FCC has allocated the 1.7–2.3 GHz band for this newer service. This shift in frequency bands opens a new door for argument and discussion. The frequencies that were selected are already in use by the fixed microwave operators in commercial, utility, and nongovernment sectors. Originally, the FCC felt that the newer spread-spectrum technology would eliminate any risk to these existing users. Early tests suggested that PCS operators would not significantly interfere with the fixed microwave users in the 1.7–2.3 GHz band. However, the tests were extremely limited in approximating the sheer number of potential users. With a sample size of only a couple of hundred, the PCS interference was negligible, appearing strictly as white noise in the band. However, if full proliferation of the PCS world takes place, the noise floor on this band will be raised significantly, thereby causing more interference.

To solve the problem, the FCC has proposed that licensees in the 1.7–2.3 GHz band be mandated to vacate the frequencies and be reassigned new frequency bands in the future. One can imagine the likely response to this argument! The existing licensees would have to be uprooted and forced into a new system. This is an expensive solution and bears heavy consequences for the current owners of equipment and licenses. To alleviate some of the financial impact for the fixed microwave users, the FCC has also proposed having new licensees in

TABLE 6.2 Frequency Band Problems for Cellular and PCS Users

Existing cellular operations	Newer PCS operations
Bandwidth already set aside	Will require establishment of new set of frequency bands
800–900 MHz frequencies not full yet, but will continue to fill up as newer services are deployed	1.7–2.3 GHz frequencies already assigned to fixed operations in microwave
No up-front cost to gain access to the frequencies	Possible $125,000–$150,000 to relocate fixed-operation microwave licensees, at a cost to the PCS operators
Limited competition, with only two operators in these frequencies in each MSA and RSA.	Possible limits and interference problems as multiple operators enter these frequencies
Not a problem for international operations, since only one organization is providing service	Not as significant a problem for international operations, since the microwave band is not as congested

the PCS arena fund the relocation of existing users. The numbers being bantered about are $125,000–$150,000 per PCS licensee, adding significantly to the cost of initially setting up the PCS systems. If PCS funding is a reality, which there is no reason to believe that it is not, then the burden is shifted in an unprecedented fashion. The existing, nettlesome issues surrounding the limited bandwidth only become magnified.

Table 6.2 summarizes some of these issues as they relate to the startup of PCS services in the United States. Obviously, the problem is less pronounced in countries where the use of private microwave is not as widespread. The government-controlled telecommunications organizations in these nations have fewer problems to wrestle with as the PCN services emerge.

Bandwidth and Channel Capacities

Much has already been said about the allocation of bandwidth for the two distinct services as they proliferate around the country. However, to ensure complete understanding, a closer comparison of the allocated bandwidth and channel capacities is in order. As noted in Table 5.4 in Chap. 5, one cellular operation being considered for the future is “sustained premium” cellular. This means that the high-end operators in voice communications will look to cellular service providers for the capacities to sustain the voice, data, and messaging needs of future mobile professionals.

The interlink may take the form of a single device or a series of

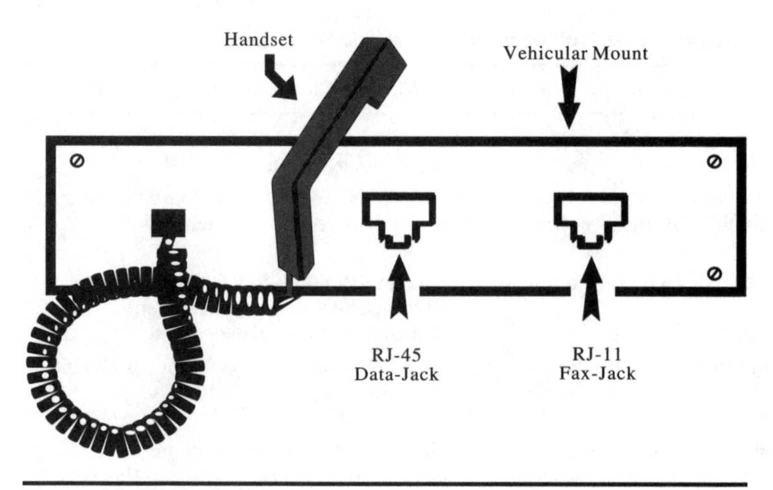

Figure 6.4 A multiconnector box can be mounted in a vehicle to allow for various services.

devices connected through a connector box. In Fig. 6.4 a connection arrangement is provided so that multiple office machines (fax, telephone, and data modem) can all be plugged into the cellular base in a vehicle. For the transportable user, this arrangement will not be convenient; some type of briefcase device will be required to interconnect the various services (see Fig. 6.5). In both cases, higher-end (standard office) equipment will be applied to the system. In a transportable arrangement, separate devices may also be used with a cellular modem communications capability. In the bandwidth for current cellular operations, the full 30 KHz channel is used to provide these services. However, this applies to analog cellular services, not to the future digital cellular services. As noted in Chap. 5, once digital services emerge in a sustained premium mode, the ability to provide a digital input will increase the range of offerings immensely.

Little has been said about the methodology underlying these offerings. Figure 6.6 shows a time division multiplexing arrangement with multiple signals being input as either digital or analog. In Fig. 6.7, the analog input from voice or fax services is converted to digital output. What now must happen is the allocation of a time slot for the pulse code modulation (PCM) data (where voice is now digital data) to move across the medium. Using the full 30 KHz of capacity, a time slot interchange will allow for the expected tenfold to fifteenfold increases in the allocated bandwidth. Higher increases are possible with a code division multiplexing technique. Adaptive differential pulse code modulation (ADPCM) uses 32 Kbps digital transport to modulate the signal onto the radio carrier. Through spread-spectrum

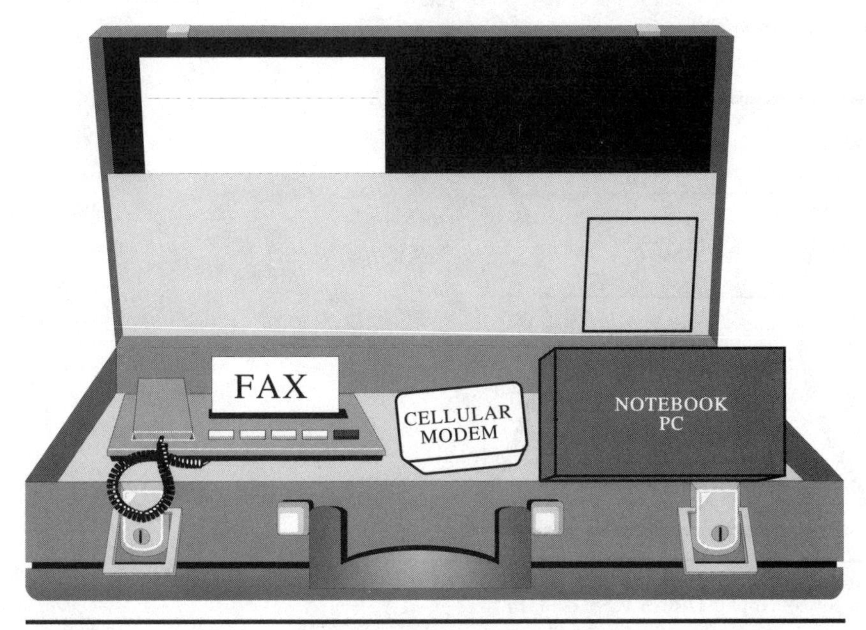

Figure 6.5 A briefcase can house multiple-device connections for the transportable user.

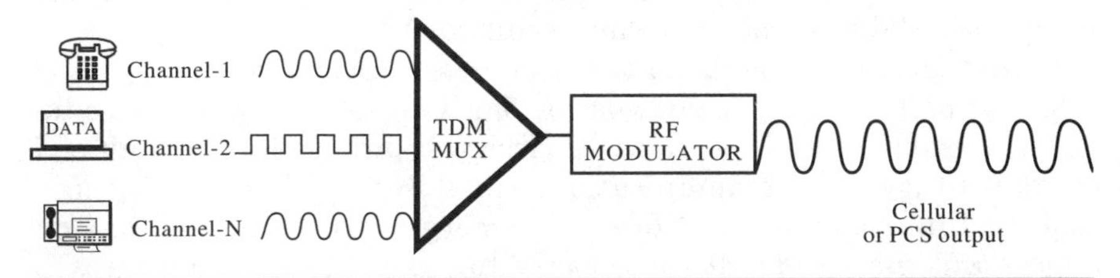

Figure 6.6 A time division multiplexing arrangement can be used for cellular or PCS output.

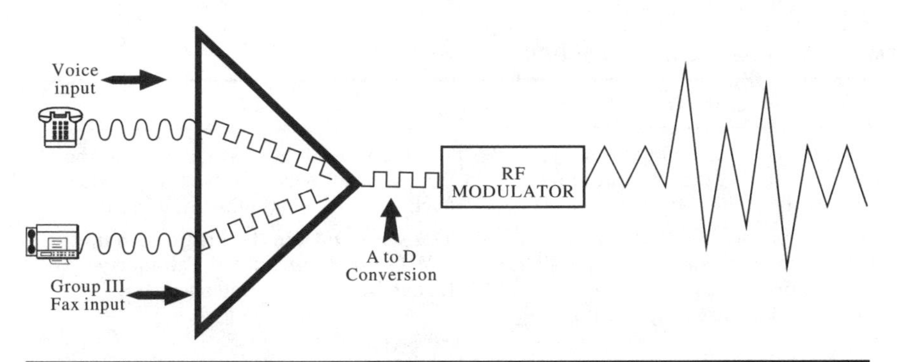

Figure 6.7 The analog input from voice or fax can be converted to a digital output for modulation onto the radio carrier.

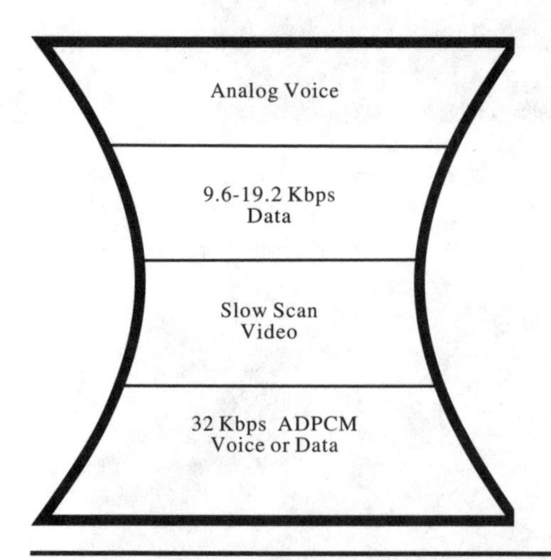

Figure 6.8 The bandwidth for CDMA will allow for 32 Kbps ADPCM voice or data.

technology, the signal will gain up to a twentyfold increase in voice or data communications on the channel. Figure 6.8 shows the allocation using the CDMA spread-spectrum technique.

Regardless of the methodology used, the transparent transport of information is the important goal. From these methods, either cellular or personal communications services will provide for the high-end or mid- to low-end communications capabilities. Microcellular and macrocellular communications will emerge as the choices for end users, and the bandwidth necessary to carry the information will become the deciding factor. In the PCS arena, a channel bandwidth of 1 MHz allocated to the spread spectrum, using either TDMA or CDMA techniques, will still serve the masses. The allocation will be

TABLE 6.3 Selecting Cellular or PCS Services

Cellular	PCS
High-end voice, fax, and data today, with opportunities to use video and multimedia in the future	Low-end voice today; data and fax in the future (by 2000); possibly other services such as video, messaging, and multimedia
Frequency division multiplexing at 30 KHz channel today; TDMA or CDMA in the future at 1 MHz channel multiplexed in spread spectrum	TDMA or CDMA in the future with up to 1 MHz per channel for simultaneous transmissions in spread spectrum
PCM	ADPCM
Analog or digital	Digital

made through various encoding techniques that use the full channel for the transport of multiuser communications. Table 6.3 compares cellular and PCS along several lines to illustrate the decision-making process of which service to select. The PCS will allow for voice communications and some limited data transport over the next decade. It seems likely that various mix-and-match services will enjoy both access methods for a business or other organization. Rather than choosing a single source for the transmission of voice and data, the user may wait to the end point to decide what price or bandwidth is going to be necessary.

In the United States, the FCC has allocated various portions of the 160 MHz bandwidth as follows:

- 30 MHz (full capacity) for the PCS major trading area (MTA) suppliers
- 20 MHz for the rural and smaller business carriers, with some allocations approaching 10 MHz for the smallest operators
- 10 MHz for the basic trading area (BTA) suppliers
- An unlicensed portion of the spectrum amounting to 20 MHz for private businesses to use for their own network services.

Ultimately, this arena may be characterized by everything from fierce competition to total disappointment with the potential offerings.

As noted, the FCC has decided to auction off PCS licenses, arguing that direct issuance of national licenses would be too expensive and time-consuming. Instead, if an existing or a potential carrier seeks a broad operating license, it must submit a sealed bid for all the coverage areas desired. If the sealed bid proves to be greater than the amount offered in the public bidding for MTA licenses, then the carrier will be granted a license in all the MTAs that were outbid. As can be imagined, this is a very progressive yet frustrating position for the potential carriers. It means that cellular operators may bid on PCS licenses outside their normal operating areas. The outcome is likely to be a consortium effort in which carriers begin to migrate their services to subsidiaries or joint ventures are formed to outgun the competition.

Table 6.4 shows the frequency spectrum planned for licenses and the possible allocation. It should be noted that the FCC has reserved a portion of the BTA licenses for preferential allocation to small rural telephone companies, small businesses, and minority-owned and female-owned businesses. This measure is designed to prevent the larger market holders from overhanging the total service area.

The FCC has also stated that future PCS operators will be required to show the financial stamina to sustain slow earnings. By 1998, only

TABLE 6.4 Allocated PCS Frequency Bands and Areas Served

Allotment (MHz)	Frequencies (GHz)	Trading area served
30	1.850–1.865; 1.930–1.945	MTA
30	1.865–1.880; 1.945–1.960	MTA
20	1.880–1.890; 1.960–1.970	BTA
10	2.130–2.135; 2.180–2.185	BTA
10	2.135–2.140; 2.185–2.190	BTA
10	2.140–2.145; 2.190–2.195	BTA
10	2.145–2.150; 2.195–2.200	BTA

a 30 percent market penetration is expected for PCS services, and over the following 5 years successful penetration will likely be 60 to 70 percent. Slow growth will, of course, stretch the financial strength of all the installers of PCS. However, the cellular services—which are also potential operators in the PCS arena—are expanding at a very rapid pace, with the past 5 years showing a 47 percent increase in customers after a very slow start in the preceding 5 years. Therefore, the market penetration projections for PCS suppliers do not seem unrealistic.

Distribution

The developers of the PCS services will still be the primary vendors or operators. However, as the market begins to proliferate new avenues of distribution will likely emerge, much as occurred in the cellular world. Early cellular suppliers used several third-party arrangements to gain their markets in various areas. Resellers of cellular services emerged and helped bring along the industry. From the user's point of view, there seemed to be hundreds of suppliers of cellular services. The fact of the matter is that only two suppliers have been allocated in each operating area. In many cases, the nonwireline companies bought out the wireline competitors in other parts of the country and created very large operating areas for themselves. For example, Mc Caw Cellular, a former cable carrier, bought up many of the licenses around the country from suppliers who were not turning a profit. This consortium approach—which helped make Mc Caw the single largest supplier of cellular service in the United States—somehow fueled the notion that there were hundreds, if not thousands, of cellular suppliers around the country.

The same approach will likely prevail in the infancy stages of PCS. A limited number of suppliers will provide service, followed by a series of commissioned resellers and remarketers in each of the operating areas. Higher-end communications services for business and

TABLE 6.5 Spectrum Allocation for PCS Services

Broadband PCS	Unlicensed PCS	Broadband PCS
1.850–1.910 GHz	1.910–1.930 GHz	1.930–1.990 GHz

government may well lend themselves to the use of aggregators, resellers, and value-added network suppliers. It will be the proliferation of newer services that will likely move business customers onto these networks.

While all this may seem like speculation, we can draw from the experience of the cellular suppliers in the past. Table 6.5 shows the spectrum that could be allocated for broadband PCS services which address the video, LAN, and other communications needs of higher-end users. It will be up to the value-added suppliers to market their services within this spectrum.

The PCS distribution business is also likely to include the cable operators. The joint efforts of the local telephone companies around the country, combined with the merger and acquisition mania of the 1990s, lends itself nicely to the notion of aggregating the assets of cable TV and telephone company operators to offer dial tone services in competition with the local telephone companies in a geographical area. Suffice it to say that the recent merger discussions between the Regional Bell Operating Companies and the cable TV companies such as Time Warner and US West Communications, and others have all left room for speculation. The cable suppliers realize that market penetration for CATV services is being realized and that newer service offerings will be necessary to gain revenue streams. To build an infrastructure of wires to the residential and business user would be both foolish and cost-prohibitive. Therefore, ventures with organizations that already understand the telephony business outside the basic service area of the partnership will be the trend. The suppliers can then aggressively pool funds, bid for licenses, and then spin off a business opportunity to deliver voice, data, LAN, and video services across the airwaves.

This value-added approach may well be the wave of the future for CATV companies, which have been under pressure to improve services while reaping very little return on investment. It also offers a new market for the local telephone companies, which until recently have been restricted from entering cable, TV, and other types of business opportunities. A joint venture or partnership arrangement with the cable operators alleviates some of the barriers. Consider the value of these mergers when the assets jointly presented show a book value of $30 billion or more. These new organizations rank quite high in the

Fortune 500 as solid investors in the technology. Their vast combined assets give them the opportunity to bid for very lucrative markets in the PCS arena.

Using the basic dial tone service, which is now a cash cow in the business, these megacarriers will have the necessary financial backing to sustain very long buildout periods until such time as consumer confidence and acceptance are achieved. Again, the rash of mergers and acquisitions in the latter part of 1993 supports this rather dramatic scenario. Further, the proposed acquisition of Mc Caw Cellular by another known entity, AT&T, suggests that the market is heating up in terms of the types of carriers that will be entering this new niche. Following the AT&T proposal, MCI announced its intentions to build out a national PCS network in support of its customer base. MCI was particularly disappointed by the FCC decision to drop a national licensing program in favor of auctioning, but the company has also stated that, if necessary, it would look to form a consortium of sorts to provide national coverage for PCS services. Sprint broke its long silence by announcing a major joint venture with Motorola for an international (or worldwide) PCS service based on Motorola's concept of Iridium (see Chap. 8). Sprint's $40 million investment into this project granted the carrier certain rights to interconnect to the global PCS network, along with some ownership rights as a prime investor (although not a major investor as far as the total funding goes).

Chapter

7

Security Concerns with Wireless Communications

Although the industry is all hyped over the use of wireless communications—whether for microwave, satellite, or cellular voice, data, or facsimile services, and now the emerging personal communications services—there remains a key issue to be dealt with: security of the information. The use of interception devices to garner information from the airwaves has long been a problem. As more and more applications are being considered for wireless transmissions, the issues will magnify. One of the prevailing reasons users have not jumped onto the wireless bandwagon more quickly is that it is difficult to secure information from unauthorized access. This access can come in various forms, many of which are undetectable. As services have proliferated in the past, the need for security has always reared its head and stagnated the process. More and more, the industry has alluded to the vast array of techniques that thieves have used to intercept or monitor the communications. If this was strictly a situation in which the telephone call was being placed over a wireless cellular call, steps could be taken to prevent the problem. Unfortunately, any wireless communications service must be protected from:

- Interception
- Monitoring
- Loss of security
- Fraud
- Theft
- Hackers penetrating networks and databases

These are but a few of the areas that must be curtailed, at all cost, if a wireless service is used.

The age-old problem of theft confronts wireless communications systems in every arena:

1. *Satellite.* Remember the theft of pay TV services such as HBO and PRISM?
2. *Microwave.* In military as well as commercial applications, information can be either jammed or copied easily.
3. *Cellular.* Consider the ham radio operator who purportedly listened in on a cellular conversation between Princess Diana and a lover.
4. *Infrared.* The airwaves are as open as in microwave, although the light is less susceptible to interception. Still, industrial espionage has been reported across a light-guided system.

These mishaps are enough to shake reality into any user or provider of services. What can be done to tighten security? The obvious answer is to use some form of scrambling or encryption to secure the information. However, the solution is not as simple as it may sound. Each of the wireless carriers has had to live through the requirements to analyze any potential entrance points into the network. Although most industry experts agree that the problems have been magnified with the introduction of cellular and now PCS services, this is not a new problem by any stretch of the imagination.

This chapter focuses on the current state of wireless technological systems and the risks that continue to shape the industry. The techniques and pitfalls that lead to great losses are examined, followed by an overview of what the carriers are attempting to do to solve the problem. This is not an all-inclusive discussion but rather an attempt to generate an awareness of the problem, describe some of the ways that information can be compromised, and provide some suggestions on what to expect from the tools that are readily available. Only extensive security training can lead to a thorough understanding of the issues. This chapter presents a smattering of the process.

The Local Exchange Carriers (LECs)

The local exchange carriers still use a lot of microwave communications in their exchange services for local and intra-LATA* calls. Approximately 40 percent of the calls are still handled by radio-based systems. This figure varies with each carrier across the country, since

---

*Intra-LATA (LOCAL ACCESS AND TRANSPORT AREA is a boundary defined on who can carry calls from area to area) calls are calls that are processed within geographic boundaries by the local exchange carriers. This definition was arrived at by the breakup of the Bell system from AT&T, and specifies the revenue-sharing arrangements for carrying calls within specific borders.

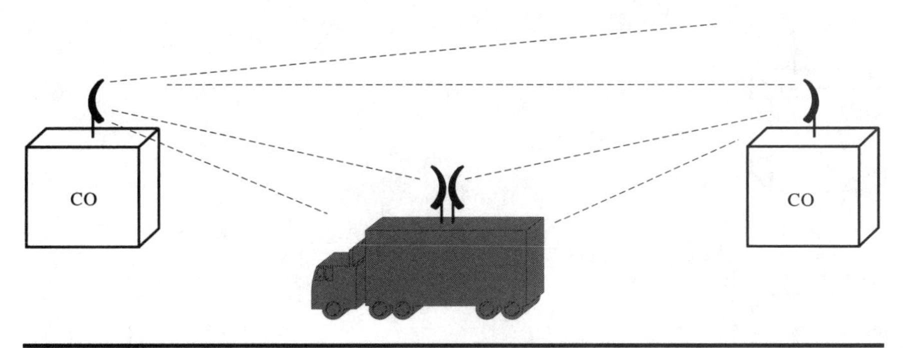

Figure 7.1 A vehicle in the path can receive the radiated energy without disrupting the connection.

there are 1465 local exchange providers comprised among the Bell operating companies and the independent operating companies. Regardless of the composition and the percentages of the calls that are carried, it takes only one penetration or security breach to cause severe discomfort to an end user.

As mentioned in Chap. 1, the frequencies used for microwave radio communications are public information, since microwave operators are licensed to operate within bounds and ranges. Figure 7.1 shows a common interface arrangement for the carrier to provide connectivity for high volumes of calls (voice, data, video, and fax) between and among central offices (COs). In this figure the systems are used to multiplex numerous simultaneous calls onto a radio frequency carrier system in lieu of the extensive wiring that would be required. The telephone companies do not guarantee security on the local dial-up calls, but they do make their best effort to secure the system whenever possible. Still, the service can be penetrated by a perpetrator by "listening in" on the conversation via radio receiver equipment. The scrambling of the calls in and among the channel allocations is an attempt to prevent such interception.

Figure 7.2 shows a radio-based system transmitting the radio carrier between two point-to-point antennas. Most people think that a

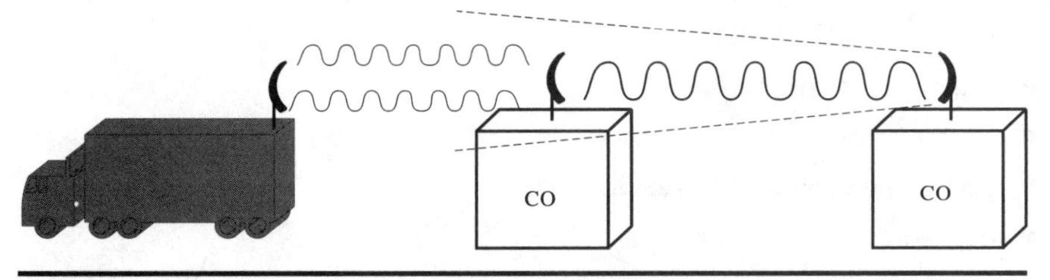

Figure 7.2 As the signal travels from exchange offices, it also continues on beyond the dish. A vehicle parked along the path could still receive the signal.

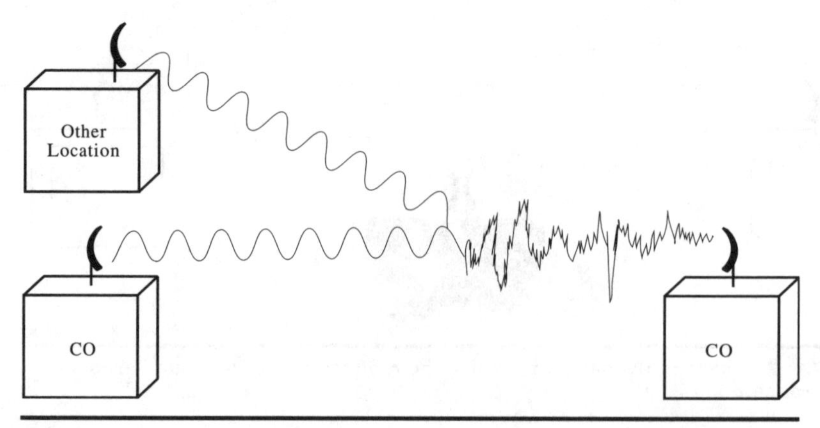

Figure 7.3 A jamming signal can cause data errors, poor quality, or total chaos.

radio signal operating in a microwave environment will only transmit between the two systems. Unfortunately, a "radio thief" needs only to be in the line-of-sight path without disrupting the signal. Since the radio wave spreads out over a greater distance, the possibilities are numerous for this interception and monitoring to take place. *Remember: Any airborne technology is insecure.* Since the thought process evolves around the reception of information, the opportunity still exists to merely jam the signal and cause the disruption at the receiving end. This could be a nuisance effort by a disgruntled individual, or a deliberate attempt to disrupt an organization's ability to conduct its everyday business.

Figure 7.3 shows how, at the local level, jamming can cause data errors, poor-quality transmission on voice calls, and total chaos on LAN traffic. Figure 7.4 addresses the types of services that may be placed across the radio system. These include:

- Voice communications
- Low-speed data communications
- High-speed data communications
- LAN-to-LAN traffic
- Facsimile transmissions
- Video-conferencing services
- Low-speed telemetry services
- Closed-circuit TV services
- Security systems

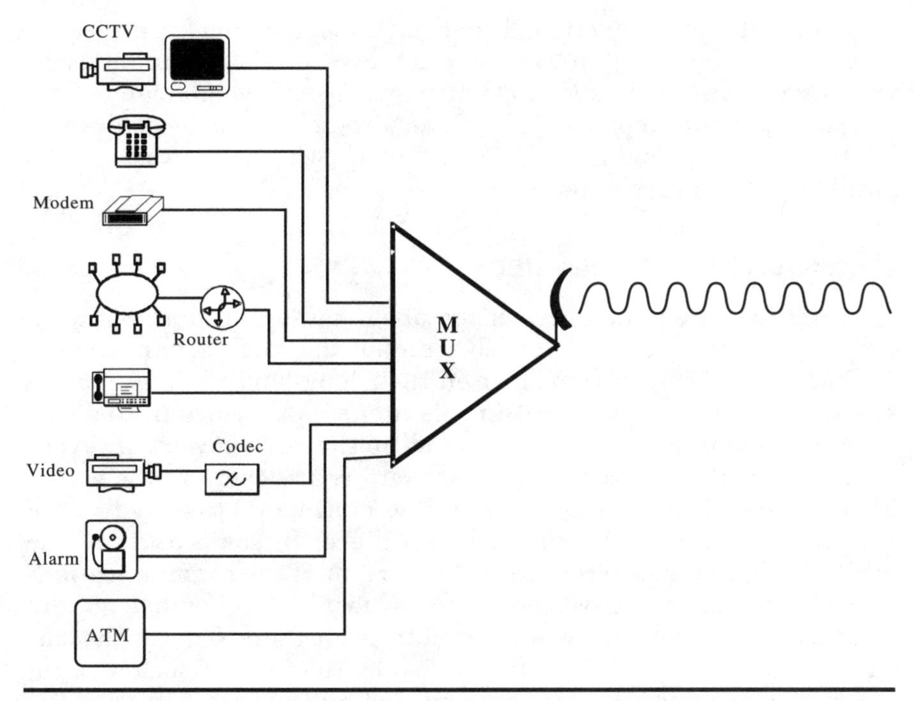

Figure 7.4 Wireless systems support a mix of voice, data, image, and telemetry services.

- Transaction-oriented services—credit card approvals, automated teller machine (ATM) access, and point-of-sale (POS) equipment
- Alarm circuits geared toward security systems and fire prevention services

As one can imagine, any interception or disruption of these services could carry significant consequences. Such services are the lifeblood of many business, government, and nonprofit organizations that one may not truly understand are running on a wireless connection. At the user end, the interface could be on a wired arrangement, such as twisted pairs.* However, once the connection is brought into the central office, it may be repackaged onto a radio-based system, without the full knowledge of the user. What is implied here is not that the local telephone companies are doing anything wrong, but that as an expedient a microwave radio system may be an integral part of the

*The use of twisted pairs or any other wired system is not to be construed as any more secure than a radio-based system. Any technology must be protected; the radio systems are merely more difficult to detect when monitoring is taking place.

communications connection. Financial considerations are also an issue here, since the carrier can achieve far more connectivity through microwave than through other means. Although many of the carriers have now implemented different systems (such as older coaxial, twisted wires, and fiber optics) in their backbone networks, radio is still a valid transport mechanism.

The Interexchange Carriers (IECs)

Although much ado has been made about the use of fiber optics in the long-distance networks, with many of the interexchange carriers boasting of 100 percent fiber on their long-haul networks, radio systems are still as prevalent in this arena as they are in the local exchange environment. Look around at the high towers in every town in the United States—for that matter, anywhere in the world. Microwave and satellite services still account for 35 to 40 percent of the total communications systems in place. In some parts of the world, radio systems account for 100 percent of the communications infrastructure. It really depends on where the call must go and what method of delivery will be used to get it there. This is the carrier's choice, unless otherwise specified by the user. If such a stipulation is requested, the carriers can and will charge extra for the connection.

For example, until the early 1980s AT&T processed one of four domestic long-distance calls across a satellite connection in order to provide enough capacity for the demands of the network users. Although this practice changed with the introduction of fiber optics, the user had no way of knowing which calls were going on a radio-based system. Further, the inherent delay on the satellite network because of the 44,600-mile trip (22,300 up and 22,300 miles back down) from a geosynchronous satellite, a delay of a quarter second each way could severely impair data transmissions. The delay was particularly true with the bisynchronous communications protocols that were prominent in this era. To overcome the problem, the user had to specify the route that could or could not be taken, for a fee.

Remember, too, that the other common carriers such as MCI* and US Sprint† are based on radio systems. Further, these organizations still must leave their networks for interconnection to regional or local exchange networks that are based on microwave radio systems.

---

*MCI originated as Microwave Communications Inc., a radio-based network.

†Originally Sprint was a subsidiary of Southern Pacific Railroad, which had microwave rights-of-way along the rail lines. Microwave radio was the basic foundation of the original network until the company went to fiber-based backbones.

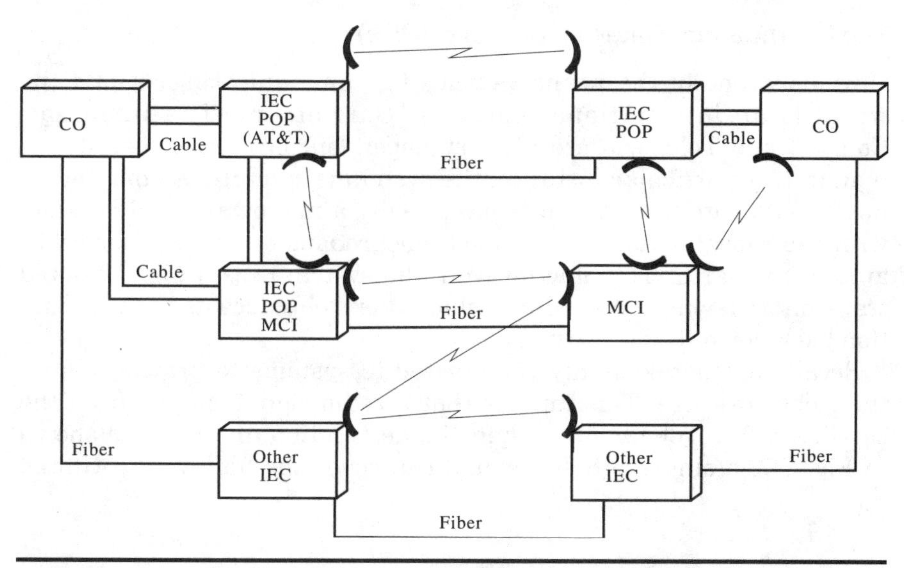

Figure 7.5 All carriers are interconnected via wire or radio systems at the local and/or long-distance level.

Figure 7.5 shows the long-haul networks via the various suppliers as a composite. In Fig. 7.6 the primary means of communications to many countries around the world mandate that the long-distance carriers use a gateway access to the international satellite-based carriers.

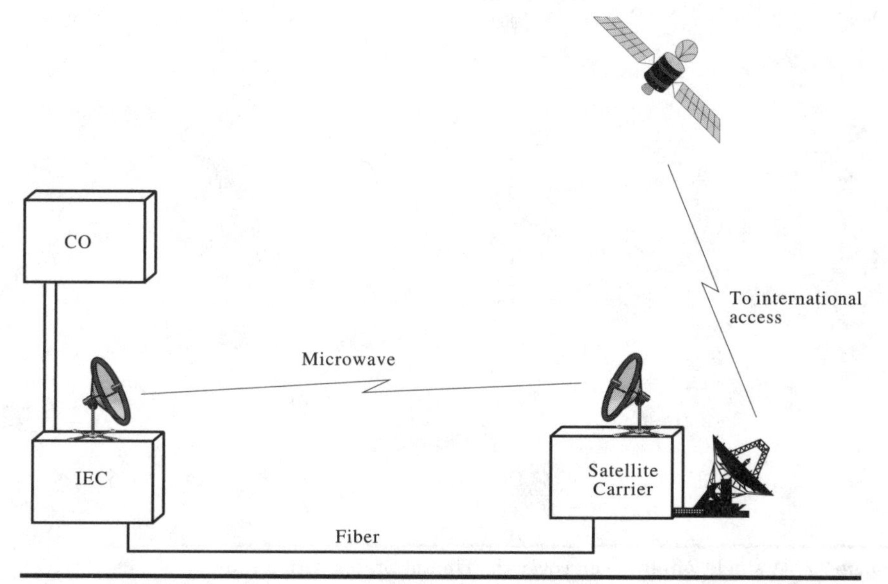

Figure 7.6 Many countries around the world have access only via radio (satellite) systems. Interconnection is therefore required at this level.

The International Gateway Carriers (IGCs)

Obviously, one of the prime vehicles for communicating around the world is an international gateway. Carriers such as Inmarsat, Comsat, and Teleports are all very dependent on a radio-based system; in this particular instance, the system is satellite. As mentioned in Chap. 2, satellite transmission is really a microwave radio system with the repeater function in a geosynchronous orbit (or other orbit) around the earth. This microwave radio system uses a high-powered transmitter facing the satellite, which then rebroadcasts the information back down to the earth.

Recall that it takes only three satellite systems to provide round-the-globe coverage. This implies that a radio signal coming from the satellite will cover far more than the destination antenna. As shown in Fig. 7.7, a single satellite signal can cover the bulk of the United

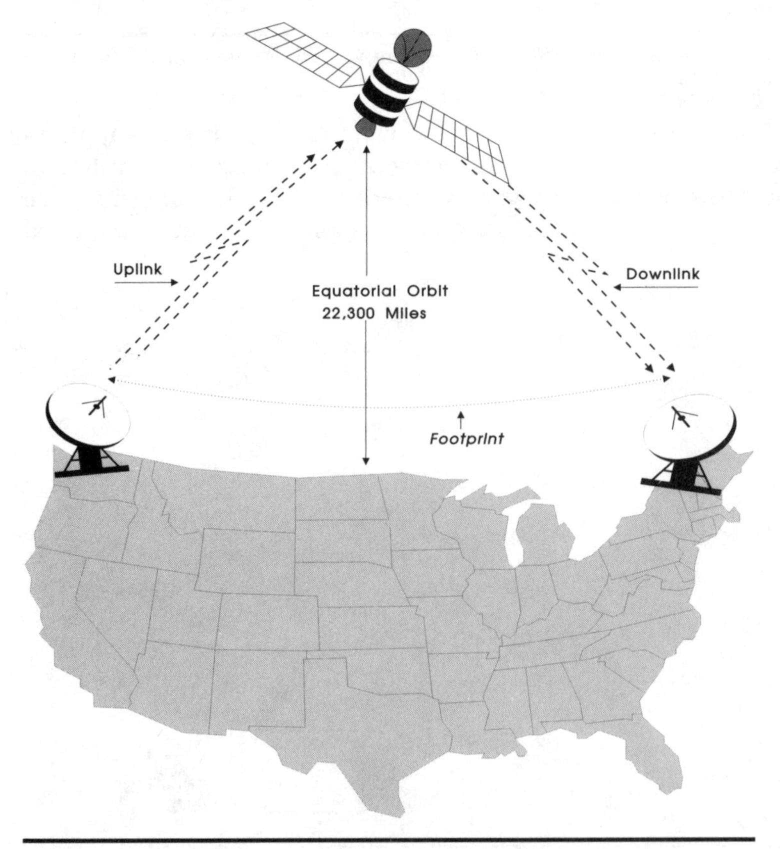

Figure 7.7 A single satellite can cover the United States with a radio signal footprint.

States. Think in terms of the signal coming to a site on the east coast but still capable of being monitored on the west coast. Now think in terms of a signal that could be monitored by a foreign location, with no possible evidence that such an event is taking place. Many of the carriers are more concerned with this interception than with jamming, since jamming distorts the signal reception. By contrast, the thieves are more interested in reception of a high-quality signal.

The Cellular Operators (COs)

What was once considered a problem for the local and long-distance carriers is now a major issue in cellular operations. The cellular industry has had significant losses from various sources since the inception of the industry. These problems are becoming far more pervasive. It is estimated that in 1992 the cellular industry lost nearly $300 million in revenues and equipment because of security breaches and thefts of various kinds. The major contributors to this loss of revenue are as follows (other areas have also been breached):

- Theft of airtime
- Equipment theft and modification
- Breaches in network security, causing loss of confidential information
- Breaches in the integrity of billing systems
- Misuse of customer database information
- Vandalism at cell sites
- Loss of customer and industry confidence

Security is becoming a major cost factor in the industry, since the cellular carriers have been diverting much of the income from limited revenue sources to tracking down these problems. The money that is being diverted is obviously better suited to network enhancements than to problem resolution. As a result, the costs of cellular service are being kept arbitrarily high, and the industry is devoting too much time to overcoming a problem that should never have existed in the first place.

Theft of airtime

One of the most disconcerting problems for cellular carriers is the theft of airtime—a situation that the industry had not predicted. Each cellular telephone has a programmed numeric assignment module (NAM) or electronic signature number (ESN). This preencoded 32-

bit address is unique to the individual set. The 32-bit sequence allows for approximately 4.3 billion individualized sets. To this numeric code the cellular carrier programs a MAN1 and MAN2 (24-bit sequence) into the set. The combination of MAN1 and MAN2 yields a 10-digit number (3-digit area code plus 7-digit telephone number). These two number sequences are also programmed into the cellular carrier's databases so that a user can be identified before a call is made.

Identifying the total sequence—the NAM plus MAN1 and MAN2—should be a monumental task for anyone and should make it virtually impossible for a thief to steal airtime. However, the cellular industry learned the fallacy of its ways when, starting in the early 1990s, breaches in the numerics began to appear regularly. Thieves have been able to re-create and manipulate NAMs to make calls without paying for them. Actually, the thefts involve the NAM plus the MAN, so that calls can be charged to the real owner. The process is good for approximately 1 month. Upon receipt of a bill, the real owner of the cellular set notifies the carrier of the excess charges and registers a complaint. The process now begins again.

Tumbling NAMs. For every measure that has ever been created to prevent a theft, a countermeasure has been automatically created to test it. Tumbling NAMs is another trick developed by thieves to steal airtime. By using a numeric sequence generator, the thieves are able to generate a valid new NAM within minutes. The use of a different NAM with each call allows the thief to charge the call to a different user each time. So much for detection and security.

Sweat or call shops. Many thieves who become proficient at cellular fraud set up locations where, for a fixed fee, users can place calls anywhere in the world without having to worry about being traced. Essentially, the sweat shop (or call shop) has a series of cellular phones programmed with tumbling-NAM capabilities. For a moderate fee (about $25) a user can walk in off the street and place an unlimited-length call to anywhere in the world without any questions being asked. In a typical call shop arrangement, a bank of telephone sets are lined up and separated by mere partitions (see Fig. 7.8). Other arrangements exist.

Theft and modification of equipment

The unique address (NAM) was also expected to be a safety valve against theft of the physical set. Every user who bought a vehicular-mounted set was assured that if the set were stolen, the industry could invalidate the NAM, thereby rendering the set useless. Every time a thief attempted to log onto the network, the NAM would be

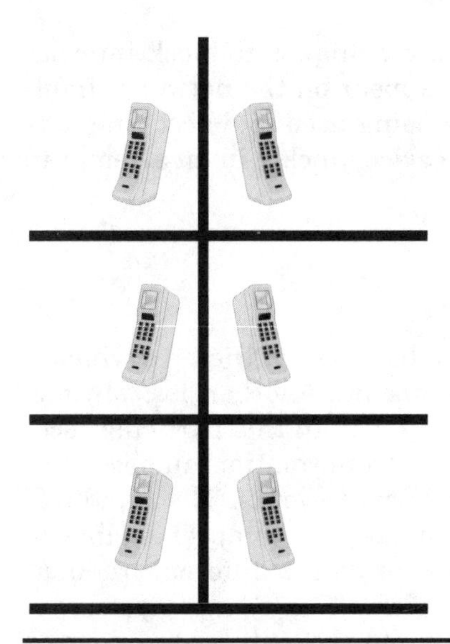

Figure 7.8 Call shops line up phones, separated by partitions. A user can walk in and place a cellular phone call to anywhere in the world for a flat fee.

rejected and the set would not work. The idea sounded good but broke down quite fast in practice. By modifying the NAM or virtually changing it, since it is on a PROM (programmable read-only memory) chip, thieves are able to reprogram the set's identity. Through a tumbling or cloning sequence, the thieves can quickly replace the NAM programmed into a stolen set with the address of a valid set. Hence, the process starts all over again. The illegal set can be used for approximately 30 days before the original owner of the address recognizes the excessive billing and reports the problem. After that, the illegal set must be modified again.

Major exposés have been aired on national TV showing how a cellular set can be changed in 1 to 2 minutes by a handheld programming device. Much of the world sat in awe when one TV network revealed the process of changing a set to a new identity (from behind a hidden camera) at the parking lot of a grocery store. The set, handed into a car window in a brown paper bag, was changed and reprogrammed in a minute and passed back out of the window. All this was done for a fee of $50 while the illegal owner unpacked groceries from a shopping cart and loaded them into the trunk of a car. Even though the modified set has a "life" of only 30 days (60 at maximum), the mere $50 programming charge yields a very high return on investment.

Sets can be modified so fast that the industry has grown quite concerned. Several carriers have instituted techniques to block international calls, or to lock in on NAMs that appear on the network simultaneously. Lastly, special programs are being used to detect irregular calling patterns and to shut down the service quickly in an attempt to hold the line on losses.

Loss of Security and Confidentiality on the Network

Theft takes on many faces. One is outright physical theft, in which a box or a set is misappropriated and disappears. Another, less obvious risk to wireless operations is shown in Fig. 7.9. In this particular scenario, a monitoring device is used to steal information, unobserved. For many users, this is an unexpected turn of events. The rapid frequency changes (cell-to-cell handoffs) make it incomprehensible to fathom how someone could still find a caller on the different frequencies. Voice patterns are usually easy to pick off on a radio-based system. Air thieves use scanners to listen in on the cellular frequencies (which they know operate in the 800–900 MHz range) and wait for

Figure 7.9 Loss of security through scanners is a major problem.

any useful information. The scanner allows the interceptor to scan through various frequencies very quickly, looking for some discussion that may prove interesting. Once interception is accomplished, the process is repeated every time the caller changes frequencies via a handoff from one cell to another. The interceptor can scan until voice recognition is achieved and then lock in on that channel and continue to follow the caller from one cell to another. Many people feel that, with the transmit and receive on separate frequencies, the interception of a call will be useless, since the party will hear only one side of the conversation. However, knowing how the systems work, interceptors can scan with two separate scanners and lock in on the paired frequency. These interceptors are not usually amateurs, but people who are really looking for usable information. When the monitoring of information is fairly straightforward, industrial espionage is facilitated.

Many corporate executives have had nightmares about what they may have said on the airwaves. Thinking that the cellular network was secure because of the rapid change of frequencies, many may have divulged information that they would prefer to take back. Whether the information was ever intercepted or monitored is unknown, but the fear is always there. The problem calls for discipline when using the cellular network: Think first.

The telephone is an instrument that everyone is accustomed to using in the office without fear of monitoring, so the security of the information is never a question. That set in the office is wired and secure (or so we think), so our natural instinct is to treat a cellular or wireless phone the same. Nothing could be further from the truth. The following precautionary steps in using wireless networks should help prevent regret in the future:

1. If you think it is safe to speak freely, think again.
2. Never say anything on the cellular or wireless network that you would not say in a crowded room.
3. If you don't want information used or listened to, don't say it. Always assume that someone is listening.
4. Never give out a credit card, bank account, or other number on a wireless set.
5. Never discuss confidential or proprietary information on any network. You may never know if a portion is on a wireless loop.

The best rule of thumb is to say only what would be repeated publicly anyway. If there is any doubt, don't say it.

Breach of integrity on billing systems

The hackers of today are getting very sophisticated in garnering information and using it to their best advantage. Imagine someone placing a call through a network to a distant country. The call lasts for an extended period—say, an hour—and is billed to the telephone number responsible for originating the call. Now imagine that this "responsible" telephone number is yours. Sounds unreal, but hackers have found ways of getting their calls billed to another person's phone. They might do so by tumbling NAMs (discussed above) or by using a cloned cellular set that mimics yours and therefore gets credited to your billing.

Another possibility is that the hackers have learned how to third-party bill. They might use a long-distance credit card number that was picked up off the airwaves. Or they might dial into the operator's network and manipulate the billing system to automatically charge a third number for calls that they make. Regardless of the methodology, the result is the same: Someone else gets the bill. Whether the billed party pays is not a matter of concern, because someone is going to absorb the cost, no matter what. If the billed third party challenges the call, the carrier will have to take it off the bill and investigate the call.

Over time, the called numbers (or countries) will emerge as a pattern. From there, the carrier will have to absorb the cost of the call, since it is not billable to the billed party. Sooner or later, the cost of these fraudulent calls will have to be forced back into the system and passed on to all users of the network services. Everyone pays. This is a large business for hackers and thieves, since it is free phone usage. It is also an opportunity for them to profit from this information, since they sell numbers to others who want to make nontraceable calls. Lo and behold the illegal trafficking expands. The industry has taken serious steps to clamp down on fraudulent billing.

Misuse of customer database information

Another problem—one that hasn't surfaced to the top of the interest pool yet—is the misuse of customer database information. Many of the wireless operators are using clearinghouses for the processing of bills and for the validation of roaming users. These databases contain all the customer's billing information, as shown in Table 7.1.

The database includes name and address (or billing address) plus information on bank account and credit cards that are used for billing purposes through the cellular operation. Can you imagine what might happen if this information were accessed by an outside party? First, the information could be used to charge purchased goods and services to this customer's card. Second, other customers' files in the database

TABLE 7.1 Typical Information from a Customer Database

Item	Description
Name	Chris Caller
Address	1412 Any Street Anytown, USA
Phone number	Business: (200) 555-1231 Residence: (200) 555-1999 Cellular: (200) 300-1234
Billing information	Bank account: 12345678 Credit cards: MC 1115551155 AX 123456789
Long-distance carrier	AT&T CC: 20055512319987 MCI CC: 20055519994357

could be manipulated so that their calls were charged to this customer's address or bank account. The possibilities play out. For obvious reasons, the database information given in Table 7.1 is disguised, and certain other information—such as the names and addresses of the clearinghouses—has been omitted.

Vandalism at cell sites

The use of cell sites along an open road, or on a dark corner of a street, invites the inevitable problem of vandalism. The cell sites are protected by fences around the equipment shack. From there the equipment shacks are protected within reason through alarm systems, various lighting techniques, and so on. However, from time to time things still happen. Vandals cause damage to the systems by throwing rocks at the cell equipment, stealing generators from the sites, or shooting through the sidewalls of the equipment shack. The list could go on forever, unfortunately. Now the added costs of securing the sites creep into the equation. Higher fences are used to deter vandals. Newer shacks are made with concrete encasements and the masts are secured more diligently. Generators are kept isolated from the main equipment, and the systems often are moved into huts that provide backup power only. Figure 7.10 shows some of these measures, all of which add to the cost of securing the systems and the networks.

Loss of customer confidence

Obviously, customers will be faced with the dilemma of what to do and how. The first reaction is that if this is such an insecure medium

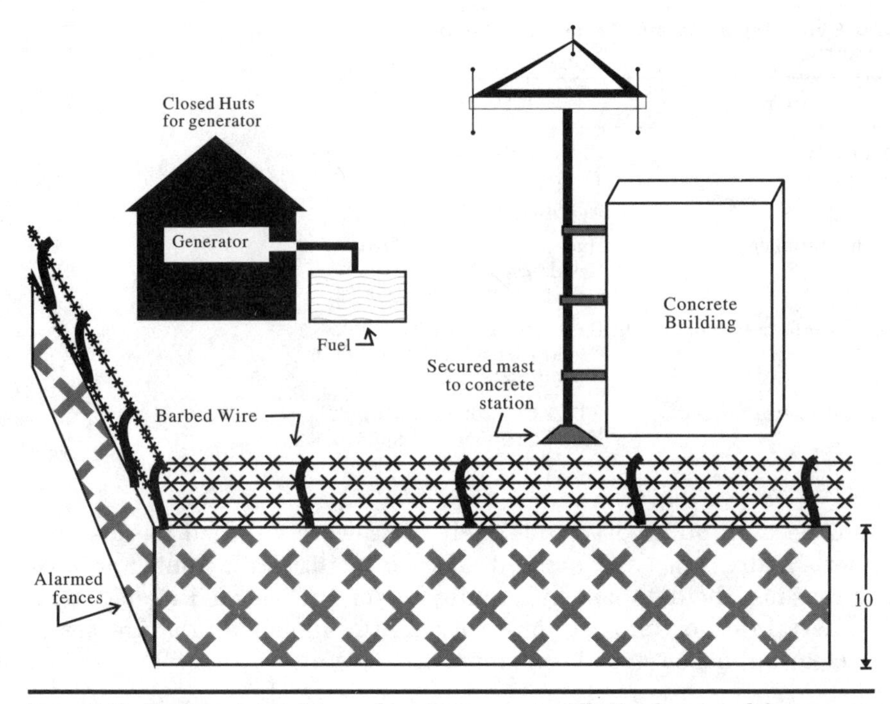

Figure 7.10 Extra costs are incurred to secure remote cell sites from vandals.

for the transmission of the organization's voice, data, and facsimile traffic, why use it? At the same time, wireless communications systems are intended to keep the traveling employee in contact with the office or with customers. Consequently, the industry has had to respond. The problem resides not just in the cellular arena but in any form of wireless communications, be it links with satellite, microwave, two-way radio, or cellular systems. Now that the industry is driving toward the PCS world, the issues and problems will be magnified.

The loss-of-confidence issue arose when cordless telephones were introduced, and users learned that anyone could monitor their conversations. Since transmission was on the open air, the FCC did not rule against the monitoring of cordless phones that operated in the unlicensed bands. However, cellular, microwave, and satellite systems operate in licensed frequency bands and are therefore protected from monitoring. Thus, the FCC declared monitoring these frequencies to be illegal. This solved the problem, right? *No, it did not stop the problem.* As a matter of fact, loss of confidence may well have increased with all the press coverage of how and what types of information were being picked off. Rank amateurs began entering the interception game. For example, media disclosures about hackers who breached the service with inexpensive scanners only encouraged others who had scanners to try it.

Combatting Fraud and Theft

The industry has had to develop an action plan to get to the root of the problem. Unfortunately, the octopus has many tentacles. At issue is not only monitoring but also theft of airtime, theft of equipment and subsequent modification, tumbling of NAMs and clones, penetration into databases, and so on. How do you mount a protection and correction plan against such a multifaceted target? Under the auspices of the Cellular Telecommunications Industry Association (CTIA), the industry has drawn from the carriers and operators the necessary funding and expertise to work with authorities to clamp down on the hardware problem. Next, an educational awareness program was instituted to alert users to the potential risk of monitoring. Finally, the industry has worked with the vendor communities to develop encryption techniques to secure information. Figure 7.11 shows a conceptual organization chart.

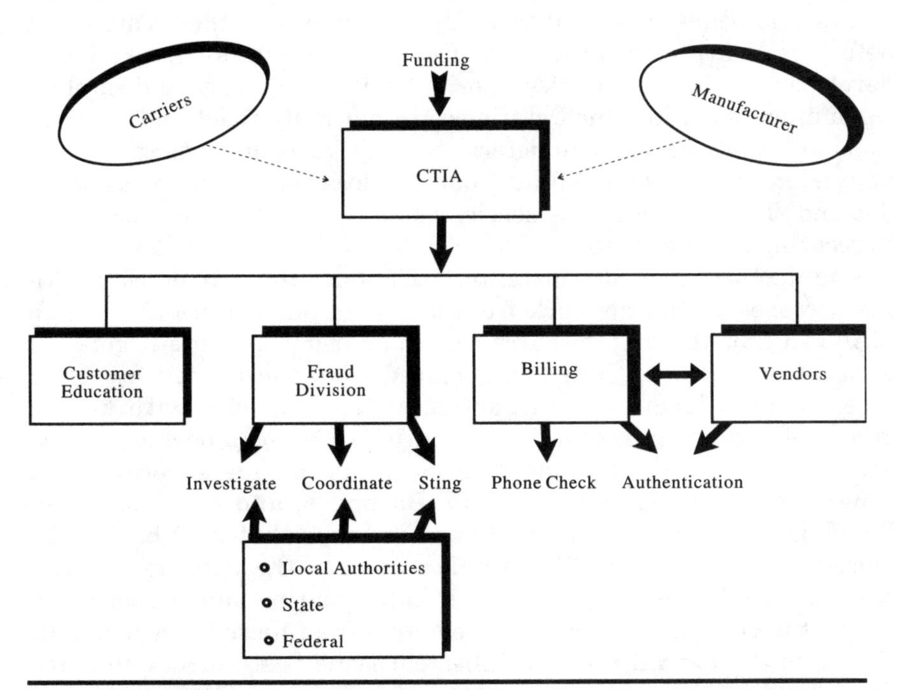

Figure 7.11 Conceptually, the CTIA orchestrates among multiple organizations to stop cellular fraud.

The CTIA

The CTIA's "fraud division" investigates theft of airtime and equipment. The division also works with local and federal authorities to track down sell shops (offering illegal equipment and services) and modification (MOD) shops. Many of the so-called sting operations have been the direct result of the fraud division's cooperation with authorities. This expensive undertaking to curtail rampant abuses siphons funds away from the normal development of features and enhancements to the networks. In 1992 the increases in theft and fraud amounted to over $300 million—a 300 percent increase over the previous year. Certainly, results are being achieved, much to the credit of the industry. However, for every win there is a minor setback, since the opposition is much longer than any task force could ever be. This adds a bittersweet note to the industry efforts.

The Manufacturers

Manufacturers are not immune to or ignoring the problem. To maintain credibility in the industry and to assist in solving these problems, cellular manufacturers are developing new methods of controlling the situation, again in conjunction with the CTIA task force. New software is being developed to verify calls and to authenticate users before calls are even processed. Encryption techniques are being developed for use in a wireless world. Unlike microwave and satellite communications, in which stations are normally fixed and destinations are relatively easy to determine, cellular operations take place at random, with users calling from any location to any location on demand. This casual-access service poses especially knotty problems in securing the information.

Manufacturers are also installing techniques that can detect multiple accesses to the network from the same identifiers (NAM and MAN) so that the access source can be shut down before it gets out across the network. Calls being placed from the United States to international locations—in particular, to countries with known groups of high offenders (such as the drug lords)—can now be blocked across the network. Troublesome locations include several South American countries, Kuwait, Iraq, Barbados, and the Dominican Republic. But what of the valid business calls that may have to be placed to these locations? Legitimate users of the network services will be inconvenienced, since their calls may be shut down in an attempt to control the problem. The outcry from these users just adds "flack" to the overall problem. The real issue is to correct and stop fraud, even if some users are inconvenienced now and then. Theft and fraud add to the cost of administering the networks and to the cost of manufacturing, keeping cellular prices arbitrarily high when they should be dropping.

Local and federal agencies

Local, state, and federal authorities have all developed task forces that work with the CTIA task force to investigate fraud and set up sting operations to catch the thieves. Recently, the New York City attorney general's office, in conjunction with the CTIA, clamped down on a sell shop netting over $400,000 worth of equipment and services. This is a monumental achievement, but victories are few and far between. Even though much more can be done, the odds are against the authorities. The world today seems to favor the crooks rather than the "good folks." Too many day-to-day issues get in the way of tracking down fraudulent users of cellular networks. The time and resources available to the coordinated task force are simply too limited.

The users

Obviously, users are the front-line defense against any scam or theft. For a start, users can work with the providers of service to set predetermined dollar limits for access to the cellular network. Therefore, it a "hit" occurs on a particular NAM and MAN, the cellular supplier can shut down access from that phone immediately. Second, patterns of calling habits are easily understood. The user can develop certain patterns or habits that will create an alert that is easily established if the pattern is deviated from. Third, user awareness of what can and cannot be put on the airwaves should be encouraged. No new user should be able to access the network without first going through an awareness program that highlights the risks associated with discussing confidential or proprietary information on the air.

As a final step, the user can keep a watchful eye on monthly bills. Verify that every time a call is billed, it is valid. Don't scan or skim through the bill; analyze it. This is an owner's responsibility, the carrier cannot do it. Whenever any form of misbilling occurs, know who to call and within what time frame. The best way to stop illegal use of the network is to catch it as soon as possible. The possibility of a misbilled call on the cellular network is there, but these are few and far between. Let the carrier know immediately if anything suspicious appears on the bill. If a data or fax transmission service is going to be used, understand the benefits of encryption and security. Talk to the supplier and to the internal audit or internal security departments before opening a door that should be secured. Table 7.2 presents a checklist to assist in using this service. These are preliminary checkpoints to using the network; more would have to be developed by the individual organization.

TABLE 7.2 Checkpoints for Allowing Users on the Wireless Network

Checkpoint	Response
1. Will the system be used for voice only?	☐ Y ☐ N
2. If the system is used for data, E-mail, or facsimile, has the user checked with internal departments before ordering the service?	☐ Y ☐ N
3. Will certain locations be called on a regular basis?	☐ Y ☐ N
(If yes, list.) ____________________ ____________________ ____________________	
4. Will other occasional sites be contacted?	☐ Y ☐ N
5. Will any international calls be made from or received by this device?	☐ Y ☐ N
6. Is there a dollar limit that should be put on the transmission of calls from this set?	☐ Y ☐ N
(If yes, $__________.)	
7. If the dollar limit is exceeded, how should the network supplier handle the service?	
Shut off immediately	☐ Y ☐ N
Call customer to verify usage	☐ Y ☐ N
Other ______________	☐ Y ☐ N
8. If an error occurs on the bill, does the customer have the local contact number and account number for the carrier?	☐ Y ☐ N
9. If a fraudulent use of the network occurs, will the customer assist in tracking the problem?	☐ Y ☐ N
10. Has the customer been advised of and agreed to the terms and conditions of use?	☐ Y ☐ N
11. Has the user been advised of the potential for "listening in" on the airwaves?	☐ Y ☐ N
12. If data or fax is to be transmitted, has the use of encryption been discussed?	☐ Y ☐ N

The carriers

Clearly, all the carriers (microwave, satellite, cellular, or other) play a critical role in stopping network theft and fraud. These carriers are funding the actions of the CTIA task force, but that just is not enough. They recognize that a three-pronged approach is best (see Fig. 7.12):

- Fund and support the task force arrangement through a user-community group.
- Educate and train users, and the industry, in what to expect.

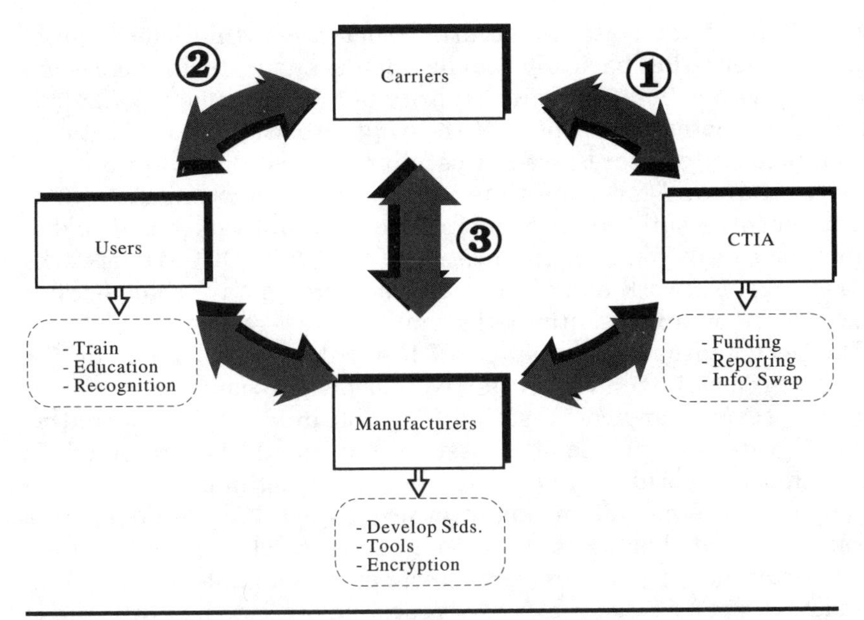

Figure 7.12 The cellular carriers are attacking the problem with a three-pronged approach.

- Work with the manufacturers to develop the necessary tools and techniques to solve this problem.

By being proactive, the carriers can encourage and support the industry, the user population, and the manufacturers. Here are some actions that carriers can take to help stem network fraud:

1. Use secure telephone units (STUs)—level III or IV.
2. Encrypt at the cell site.
3. Encrypt at the set through an external box.
4. Encrypt at the MTSO.
5. Use multiple levels of encryption.
6. Employ trap tools to detect fraudulent use of the network.
7. Make the move to digital cellular quickly.
8. Be proactive in chasing offenders and prosecuting them to the fullest.

The single largest risk of encryption techniques is building a false sense of security, leading users to neglect the safeguards of the system as it is designed. In many cases, the system is secure; it is the user's failure to activate the security (encryption) that causes problems. Further, a lull basically sets in when users deal with nonsecure callers or with callers who have security but do not announce it. This lull defeats the entire purpose of securing airborne communications. Again, security measures are not peculiar to the cellular world. They have been around for some time in the other communications networks, such as satellite and microwave. The military began using automatic secure voice communications (AUTOSEVOCOM) networks decades ago. Network names and service offerings have changed, but the concept remains essentially the same.

The best approach is to start off the right way by securing (or encrypting) a call no matter how trivial or insignificant the conversation may seem. Many conversations start off in idle business chatter then migrate to very sensitive issues that should be protected. It seems that the world is full of people who have nothing else to do but listen in on the conversations of others. Not all these people are crooks or thieves hoping to learn something secret or private. Many are just curious types who have the technology available to them. It is the more serious eavesdropper who poses the biggest risk. But who is to know if a serious eavesdropper is on the airwaves at the same time? The gamble is too great to take lightly.

The Encryption Process

Many of the carriers have developed techniques to encrypt at various points along the network. In a cellular world, encryption begins at the cell site (see Fig. 7.13). Remember that the individual set on the cellular network listens to the paging channels and waits for its address to be called out. The network then allows the set to perform the call setup (and breakdown) on the paging channels. Therefore, the signal is in the clear (unencrypted) until it gets to the cell site itself. From the cell site out to the rest of the network the carrier can offer the encryption service on a call-by-call basis, for a fee. Unfortunately, to deliver the call to the sets on both ends in this environment, the network switches must do the encryption and decryption, leaving the "last mile" unsecured. This really defeats the whole purpose of the service.

Key systems

Fig. 7.14 introduces a slight modification to the overall process. Here an encryption device is installed inside the vehicular cellular telephone in the form of a hardware chip and some software algorithms to provide the appropriate levels of encryption. Of course, the user on

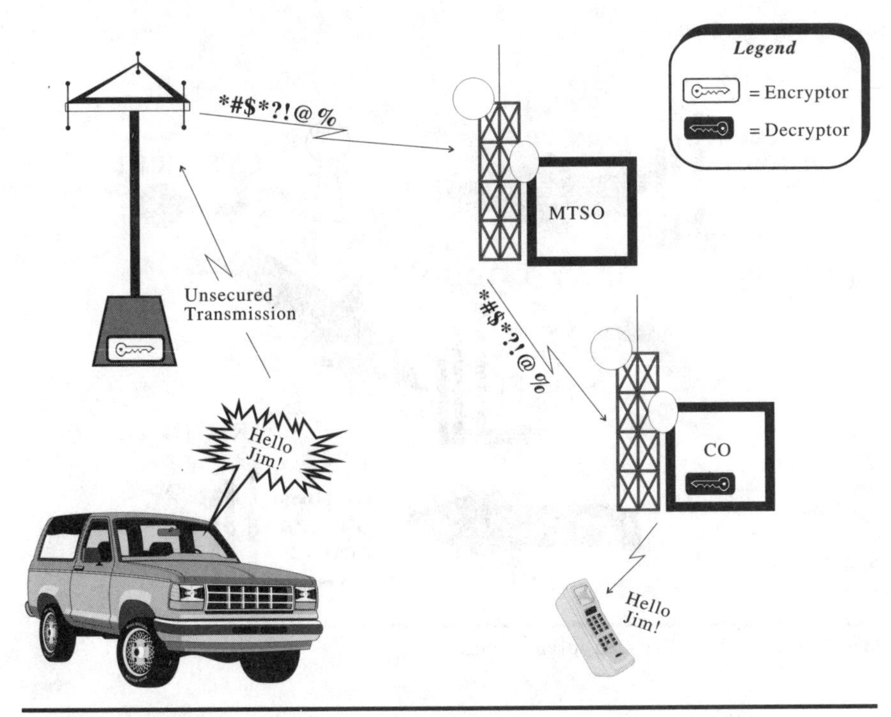

Figure 7.13 Encryption at the cell site is a first step.

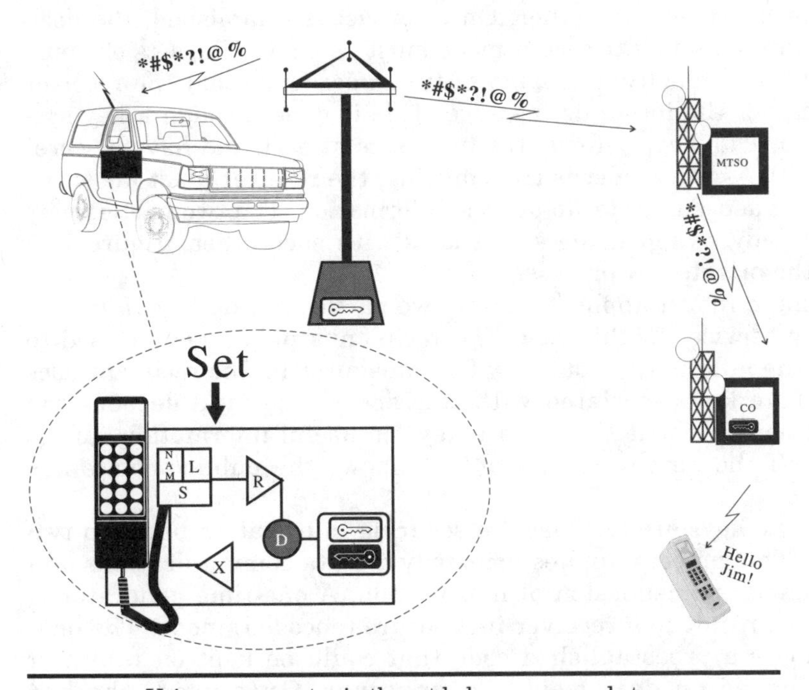

Figure 7.14 Using an encryptor in the set helps as a second step.

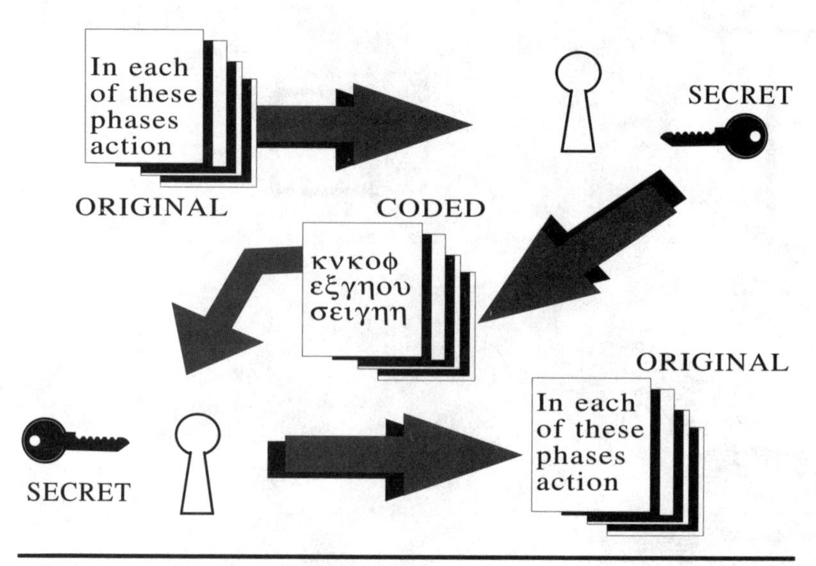

Figure 7.15 A private key encryption technique.

the other end of the call must also have an encryption capability or the system is for naught. The two parties set up the initial connection and converse with each other. Once contact is established, the decision to flip the sets into secure mode must be made. The sets obviously must have some commonality of transmission so that the receiver can decipher the incoming message. This is done through a key system. In *private* key systems, the two users at each end have a secret key (code). As the sender is transmitting, the receiver must know the decode sequence in order to use the information. The two parties may have already swapped the secret keys with each other. Figure 7.15 shows the private key process.

In a more robust application, the two parties can use a *public/private* key service. In this case, the recipient's public key is used to encode the information, but only for transmission. The recipient uses the private key associated with this specific set and decodes the information. Without the private key, no useful information can be gained off the airwaves. Figure 7.16 shows the public/private key sequence.

One-time keys are also used in securing information between two parties. This process applies primarily to data communications and the financial transmission of information. A one-time generator is used to transmit to a receiver in a secure (encoded) mode. The one-time key is a preestablished code that could be kept on tables or charts based on day, week, or whatever. Once used, the key

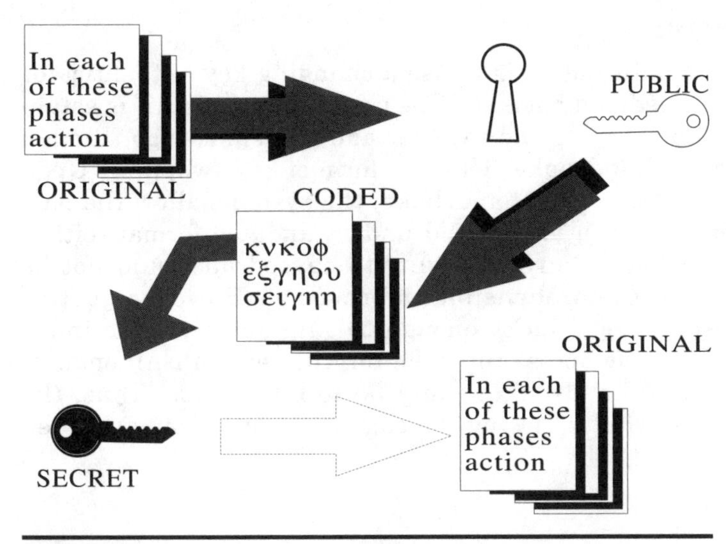

Figure 7.16 An alternative to private encryption is the public key system.

becomes invalid and a new sequence is selected. Although the technique has been used for many military systems in the past, it is also being applied to the transfer of financial funds. For example, the banking communities use SWFTE to perform international electronic funds transfers. Some one-time keys have a means of generating a false key in the event that eavesdroppers are listening in. In this case, the key may have two meanings, as shown in Figure 7.17.

ENCRYPTED MESSAGE:	W	C	N	V	G	Z	M	S	P	E	Y
KEY#1	4	14	19	8	3	7	6	1	11	4	5
SECRET MESSAGE:	S	O	U	N	D	S	G	R	E	A	T

ENCRYPTED MESSAGE:	W	C	N	V	G	Z	M	S	P	E	Y
KEY#2	9	24	9	18	14	6	21	14	14	19	18
SECRET MESSAGE:	N	E	E	D	S	T	R	E	B	L	E

Figure 7.17 One-time keys can produce a different result for eavesdroppers.

Authentication systems

Another security alternative is to use a changing key authenticator on a regularly scheduled basis. In this case, when a circuit is established across the network, the receiver and transmitter go through an authenticating handshake. This is a form of key swapping. Upon successful completion of the handshake, the systems allow the passage of information in a usable and understandable format (either voice or data). Clearly, anyone trying to eavesdrop would not be able to use this incomprehensible information. However, just in case the eavesdropper is lucky enough to gain access to the information as it is moving across the medium, the sets on both ends go through a series of handshakes every 30 to 60 seconds. Thus, the eavesdropper will wind up losing the edge and receive only encoded information.

Multilevel encryption

Multilevel encryption is an added technique currently being experimented with by some carriers. In this case, depending on the system, the carriers will use a public/private key arrangement between the cell site and the user's set. From there the cell site will reencrypt the information for transmission across the network. If the last use in on an encrypted set, the information will arrive at the set in an encrypted form. However, if the called party (or the caller, depending on the actual sequence) is not on an encryption device, the last connection (either the MTSO or the CO) will decrypt the information and deliver it to the final destination. Each encryption or decryption step along the way will increase the latency across the network, potentially adding a total of 30 ms to the transmission process. Users must be aware of just where the steps are taking place and where the information is in an unsecured mode. Figure 7.18 illustrates the multilevel encryption process.

The costs of all these services are the variables in the equation. Some carriers are charging monthly for the services, raising the average monthly service access costs to $90. One-time fees for the encryptors and decryptors are in the range of $300–$1000 per end.

Questioning the carrier about encryption services

Questions must be asked that will address the overall understanding of the service and where it takes place. Here are some of the questions that might well be asked of carriers:

1. Where does the encryption actually take place?

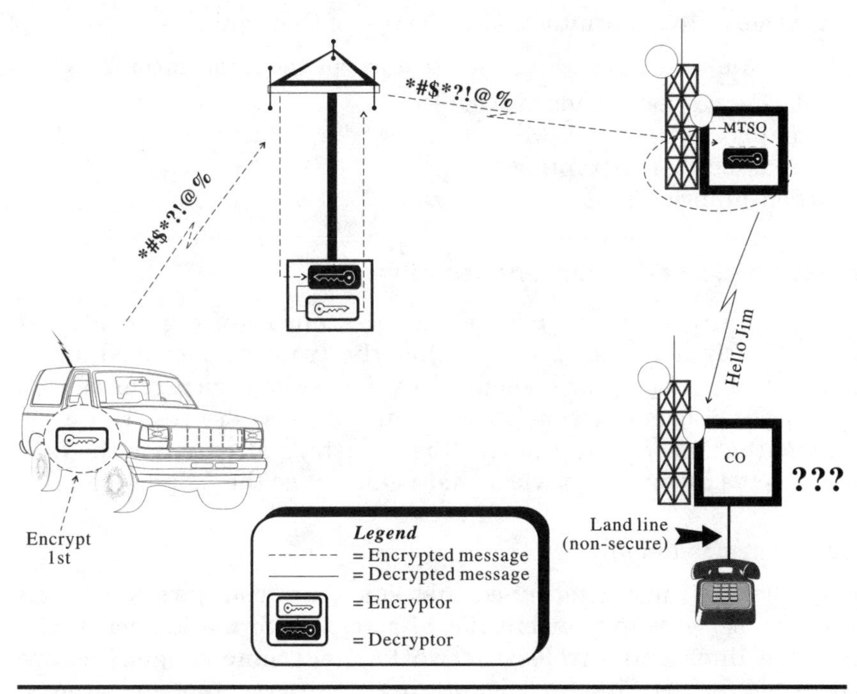

Figure 7.18 Multilevel encryption is better, but latency increases exponentially.

2. What form of service is provided?
   Private key?
   Private/public key?
   One-time key?
3. Are multiple levels of encryption offered?
4. How much of the transmission path is in the clear?
5. What if the user on the other end doesn't have secure equipment?
6. How does the service get activated?
   Every call?
   Toggle switch?
   Other?
7. Are authentication methods used on the network?
8. What benefit will digital transmission offer in securing the network?
9. How much is the service going to add to the monthly billing?
   Call by call?
   Monthly service increase?

10. Are there any guarantees with the use of this equipment?
11. What standards are being used to encrypt the information?
    Data encryption standard (DES)
    Semaphore
    RSA associates encryption
    Other, proprietary

Issues That Could Circumvent Security

Security is always a hot topic when the transfer of organizational information is involved. The fact that the transmission medium is unsecured should provide enough of a wake-up call to current as well as prospective users of wireless communications services. However, the issue that has recently caught the industry's attention is the U.S. government's involvement with what is known as the "clipper chip."

Government monitoring

The government has determined that since the "bad guys" are using encryption services to transmit illegal-activity information across the telephone lines and wireless networks, government agencies are responsible for tracking and monitoring these activities and groups. When government agencies gain permission through warrants to listen in on (wiretap) the circuits, the use of encryption stymies their efforts. Therefore, two copies of all new keys (the coded sequence and a copy of the computer chips that perform the encryption) created for encryption services must be sent to the government.

Obviously, there are two sides to this issue and the debate will rage for years to come. The point is that wiretapping and eavesdropping are on the rise, both from government agencies and from hackers and electronic thieves. Again, an alarm should sound to business organizations that there may be no such thing as security in communications. If the government is arbitrarily listening in on conversations and monitoring data transfers of corporate information, of what value is that information? This is not a statement against the needs of government agencies to capture and prosecute drug dealers, felons, and other criminal elements. Rather, it is an alert to people who have no idea of what is going on that they must become far more aware of current circumstances. If agencies are listening in on business conversations, what are they hearing? What corporate direction statements or private information is going across the airwaves in an unsecured manner?

The story of two senior executives who transmitted what was thought to be confidential information across the cellular network sums up the current environment. Apparently, what these two executives were discussing was construed by government agencies to be collusion and price fixing. Therefore, the channels were monitored as a matter of course, without the knowledge of the executives in question. However, government agency personnel heard enough across the airwaves to obtain an arrest warrant and ultimately to prosecute. Was the information legally obtained? Are there exceptions to the monitoring rules? Can anyone just decide to listen in on conversations without jeopardizing due process?

Where does it all start? Who are the friendlies and who are the foes? Again, this statement is directed, not at the agencies in question, but more at the implications of the total process. If the two executives were wrong in their actions, then by all means it is the responsibility of government agencies to obtain the necessary information to prove their case. The overriding issue is how that information is obtained. Forewarned is forearmed. Further, if the government is given copies of all manufactured codes being used to secure information through the "clipper chip," who else might be able to obtain copies of the same list of keys in use?

Digital migration

Digital transmission may offer some relief to the general problem of eavesdropping. Many of the carriers are reluctant to make major investments in encryption equipment because of the costs, the legal consequences if breaches occur, and the industry's plan to migrate into a digital transmission service that could well negate the need for encryptors.

How will digital help? First, with direct or digital sequencing, the information will be mixed and scrambled onto the airwaves in a different form than it is today. When specific frequencies are used and the entire channel is allocated to a single user (the current FDMA system), eavesdroppers can readily listen in. However, if multiple conversations are digitally encoded and multiplexed onto the same channel but in time slot interchanges (TDMA), the process becomes extremely difficult to follow with the state-of-the-art scanners on the market today. Alternatively, if a 10-bit coded-chip sequence (CDMA) is spread across multiple frequencies through a frequency-hopping scheme, listeners may not be able to track the information. Or the coded chip may simply be difficult to obtain and decode.

Figure 7.19 shows digitally encoded information in a TDMA sequence. As can be seen, if the digital sequencing begins at the set, it will not be easily listened to across the radio waves. Figure 7.20 shows the frequency-hopping CDMA spread-spectrum sequence as a

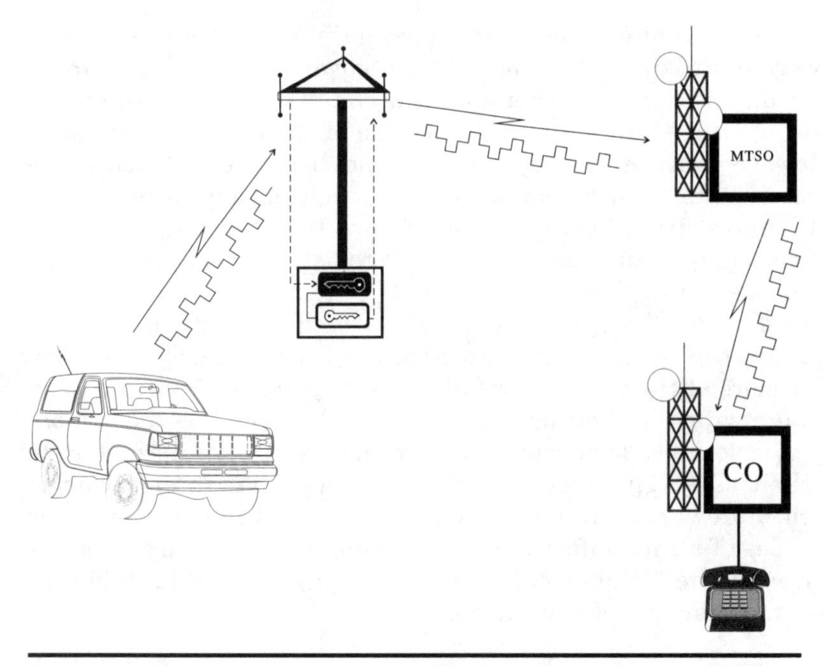

Figure 7.19 Digital transmission with TDMA will help enhance security.

means of proving this point. How would a single scanner be able to pick off the information and reassemble it into a logical sequence? Even if multiple scanners were used, the reestablishment of the sequence, the frequency-hopping scheme, and the final coding would be a formidable task for any eavesdropper.

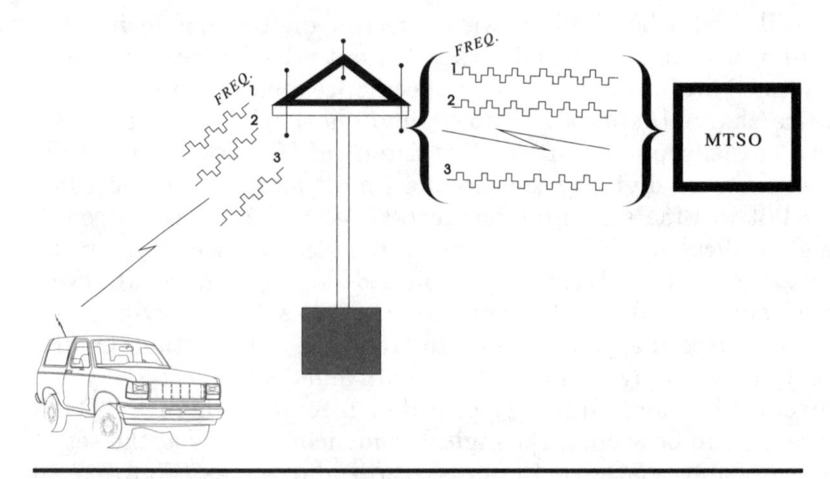

Figure 7.20 Frequency hopping and coded-chip CDMA further improve security.

This is not to say that professionals with extensive equipment will not be able to break the sequence. However, the fact that a different sequence will be used with every transmission complicates the process exponentially. Over time, the systems will likely be broken. For every measure that is created to solve a problem an instant countermeasure appears to test the process. Unfortunately, as discussed in previous chapters, the migration to digital, plus the decision to use TDMA or CDMA, gets in the way of full deployment. Therefore, the digital systems are not likely to be ubiquitous until the late 1990s or early 2000. Users may have to use a temporary encryption service until such time as the digital systems are readily available.

Chapter

8

VSATs, VSATs, VSATs, and More

Quite a bit of discussion has already been afforded the current state-of-the-art wireless technologies. However, two separate technologies are merging to shape a new service for the future. These are long-haul communications and personal communications. The new service brings together past (satellite) and future (cellular) concepts to offer the promise of instant worldwide communications.

Satellite-based service will use very small aperture terminal (VSAT) or spot-beam transmission as the vehicle to deliver global connectivity. This technology brings home the concept of communications from anywhere to anywhere. The thought of being out in the middle of a lake and receiving a call, or skiing down a slope and making a call, boggles the mind. There are many rural or undeveloped areas in the world where no telephone service infrastructure exists. In a matter of a few years these remote locations—in forests or jungles, on mountains or seas—will all be reachable within a moment's notice. It is important to look closely at this new combined service to understand just what is happening in the global arena.

Invasion or Innovation?

There are some people who feel that personal communications services will destroy the last remnants of privacy and personal life. On the other hand, there are superworkers who feel the need to be in constant communications reach of their office at a moment's notice. The infrastructure of a wired world, with its barriers of timing and cost, does not lend itself to instant communications. A wireless transmission system is the obvious answer. However, cellular or personal communications technology alone still leaves a lot to be desired. First, this type of service is always going to be deployed initially in major metropolitan areas, where the usage rates and financial returns will be achieved. It will be decades, if not longer, before the technology ever works its way close to the remote areas. Even then there will be geographical pockets that never get the service, since the installation, maintenance, and coverage costs would be prohibitive.

Enter the ability to see the world from above! If forests posed no problem with the need to clear out dense terrain for communications huts; if mountains offered no major challenge to scale the heights of sheer rock; and if bodies of water were not in the way of establishing a relay link across major portions of the earth—then *all* the communications services could become ubiquitous and cost-efficient. Thus the industry decided to attempt servicing these areas from a satellite capacity. What is new is not the concept—satellite transmission systems have been around for over 30 years—but rather its application to an on-demand dial-up service.

It is exhilarating to think that new life could be breathed into old technological ideas and that new services are mere extensions of the old. It is poetic to suggest that the future relies on the past as the communications industries around the world redeploy older systems with a new marketing approach. The venture will have to be lucrative, because the costs will still be quite high. At present, over a dozen suppliers are competing for space segment and frequency allocations to offer voice, data, paging, and radio determination position services, as a minimum, in the future. These organizations are considering various approaches and orbital capabilities for launching their service offerings. The most widely discussed approach today is *low-earth orbits* (LEOs) communications. However, *mid-earth orbit* (MEO) and *geosynchronous orbit* (GEO) satellite-based communications also merit attention.

The various organizations applying for the rights to use frequencies are shown in Table 8.1. The experimental licenses are being discussed at great length around the world, especially in the United States, as the future interoperable service. An immense amount of activity is going on in this industry, and competition for the allocation of limited spectrum is fierce. The drum beats to the tune of various drummers. Who will emerge as the winner will be decided over the next 7 to 10 years.

Regardless of the orbit used, it seems likely that the first entrant in the global satellite business will gain the market share. All others will have to work harder to steal or maneuver market share away from the industry leader. This is an expensive proposition, especially if some other carrier gets there first and takes the market by force. Of course, some observers feel that there is no need for these systems to begin with, and that the competitors will go bankrupt in the first few years. Be that as it may, industry shakeouts do occur. The jockeying for market position remains an important factor.

TABLE 8.1 Satellite-Based Competitors and the Orbits Being Sought

Orbit	Number of competitors	Status
Low-earth orbit (LEO)	5	Pending, experimental licenses granted for very specific areas of coverage
Mid-earth orbit (MEO)	4	Experimental licenses granted for specific areas of coverage
Geosynchronous orbit (GEO)	4	Licenses already issued for some; experimental for others.

Low-Earth Orbiting: The IRIDIUM Concept

In December 1990, Motorola Satellite Communications Inc., a subsidiary of Motorola Inc., filed an application with the FCC to construct, launch, and operate a low-earth orbit global mobile satellite system known under the trademark IRIDIUM system. This was the hot button that sparked the world into a frenzy. The IRIDIUM concept involved launching a fleet of satellites around the world to provide global coverage for a mobile communications service operating in the 1.6100–1.6265 GHz bands. Motorola proposed using a hand-held portable or mobile transceiver with low-profile antennas to reach a constellation of 66 satellites.* The satellites would be interconnected through a radio communications system as they traversed at 420 nautical miles above the earth in multiple polar orbits. The system would provide continuous coverage from any point on the globe to virtually any other point on the globe. The use of spot-beam technology, which has been discussed for years in the satellite industry, allows far greater frequency reuse capacities than have been achieved before. Iridium, Inc. (an international consortium of telecommunication and industrial companies [spinoff of Motorola] funding the development of the IRIDIUM system) has proposed four types of services, each of which can be provided in any of three coverage areas—air, land, or water. The services are:

- Voice communications
- Data communications
- Paging
- Radio determination

---

*The original concept was to use 7 polar orbits with 11 satellites in each, similar to an orange-slice concept, to provide worldwide coverage. The 77-satellite concept was amended to 66 after the World Administrative Radio Council meeting in the spring of 1992.

As the primary entrant in this arena, Motorola also suggested that an interconnection arrangement be set up with local post telephone and telegraph organizations (PTTs) around the world. The concept is sound and, as described in detail below, the approach will provide for the coverage to remote areas that has been lacking in the past. The IRIDIUM satellites will be relatively small compared with others that have been used. The electronics inside will be very sophisticated, and System Control Facilities on earth will provide the command and control service for the administration of the overall network.

As much as the cellular networks have prospered and proliferated on land, there are areas of coverage that cannot be achieved. The cellular networks have used a type of spot-beam coverage through directionalized antennas. However, the cells are fixed in specific sites; the user is the mobile target being handed off from one set of cells to another. Also, even though the cellular networks can accommodate the interconnection of various carriers, the user must be within a geographic area that provides the service.

The IRIDIUM Network effectively reverses and expands the cellular concept. The IRIDIUM cell sites are constantly moving above the user population, providing coverage in areas where cell sites would not be practical. The user remains relatively fixed. Table 8.2 compares the cellular and IRIDIUM concepts.

The cellular networks are primarily located in the United States—across 1500 licensees covering that many geographical areas. However, there are still a lot of areas where it does not make financial or operational sense to install service. As a result, very rural areas, mountain terrain, and other locations do not receive coverage at all. Under this concept, all the areas not previously served will be easily accommodated. The IRIDIUM Network will even provide coverage in the extreme rural areas of Alaska. The IRIDIUM satellites will address not only obvious services—such as two-way voice communica-

TABLE 8.2 Cellular Versus IRIDIUM Networks

Cellular	IRIDIUM Network
Sites are fixed.	Sites are the moving targets.
User moves from site to site.	User stays put; sites move the user from satellite to satellite.
Areas of coverage are 3–5 miles across.	Areas of coverage are 185–1100 miles across.
Coverage is sporadic, not totally ubiquitous.	Coverage is worldwide.

TABLE 8.3 Initial Features of the IRIDIUM Network

Feature	Description
Radio determination services (RDSS)	Location of vehicles, fleets, aircraft, and marine vehicles. The RDS system (RDSS) will also be an integral locator service for all voice communications devices.
Voice communications (VC)	On-demand dial-up digital voice communications at 4800 Bps from anywhere in the world
Paging (P)	A one-way paging service that is fairly well established. The paging service will include an alphanumeric display for up to two lines of characters for global paging access instantaneously.
Facsimile (fax)	A two-way facsimile service capable of transmission speeds of up to 2400 Bps.
Data communications (DC)	An add-on device to the voice communicator allowing for the transmission of two-way data at speeds of up to 2400 Bps. This capability will also allow for two-way messaging (E-mail) service across the network.

tions—but also other services that may not be as evident. Table 8.3 summarizes the initial features of the IRIDIUM system.

The LEO arrangement underlying the IRIDIUM concept is shown in Fig. 8.1. In this particular case, the satellites are moving in a polar orbit at more than 400 nautical miles above the earth. Recall from Chap. 2 that there are three types of satellite orbits: polar, elliptical, and equatorial. In the polar orbit, the satellite traverses the poles of the earth's surface and therefore passes over any specific point along its path very quickly. The satellites will move at approximately 7400 meters per second. Therefore, as one target sites moves out of view, a new one will come into view at approximately the same time. The handoff will take place between the individual satellites using the K_a-band of frequencies.

Motorola estimates that 200 MHz of bandwidth will be required for the intercommunications service between satellites. Figure 8.2 shows the ground telemetry and control services, known as *gateway feeder links*. These will also use the K_a spectrum at approximately 200 MHz of bandwidth. Initially, the IRIDIUM network will need to use approximately 5.1 MHz of bandwidth in the L-band. However, Motorola anticipates that over time this will increase substantially. The L-band will be used for the communications capabilities from the

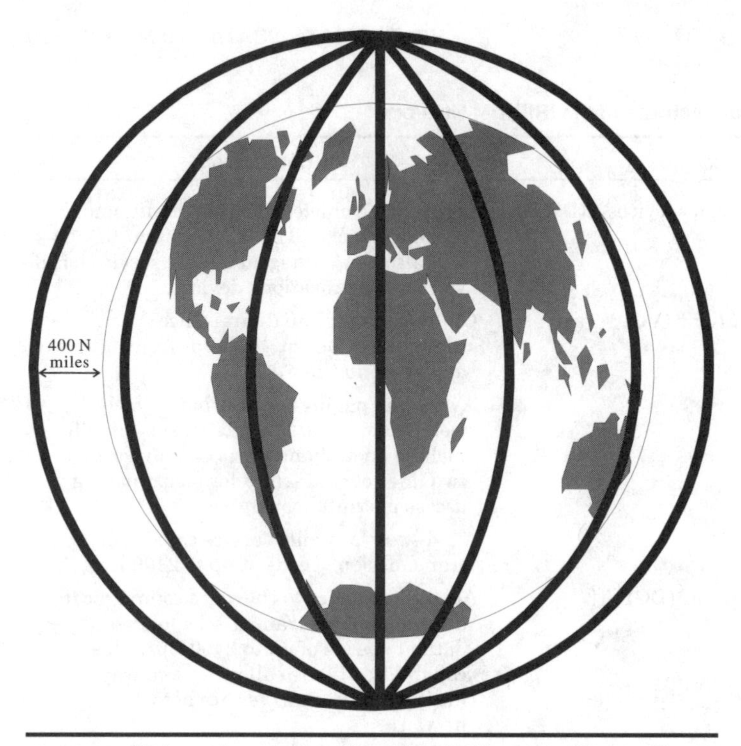

Figure 8.1 Using a low-earth polar orbit, Iridium satellites will move at about 420 nautical miles above the earth's surface.

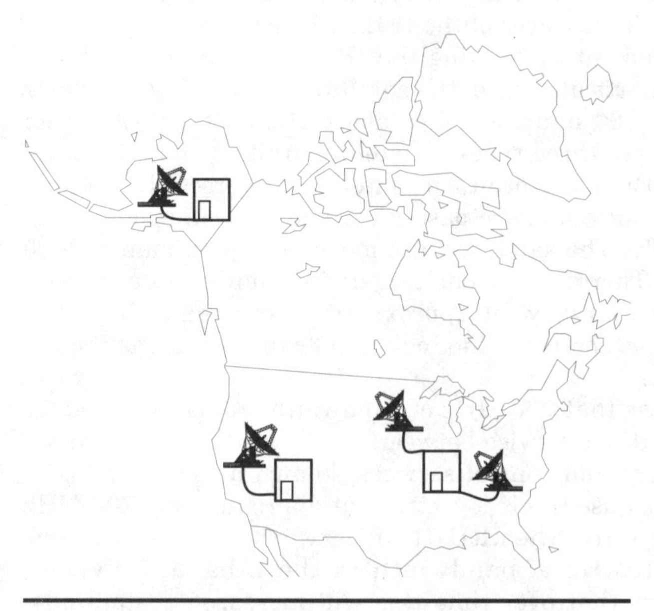

Figure 8.2 Command and control is through gateway feeder links from ground satellite stations.

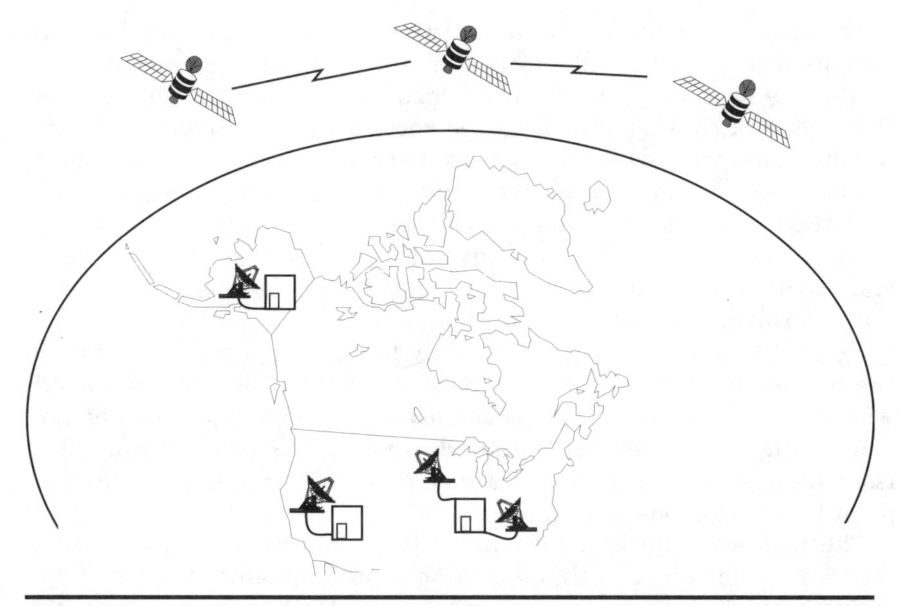

Figure 8.3 Satellite-to-satellite handoff will use the K_a band frequencies.

handset to the satellite, as shown in Fig. 8.3. A subscriber can communicate directly with the satellite, or cell, in this environment.

The L-band is a special band reserved for radio determination services (RDSS) and ancillary messaging satellite services. Therefore, in order to gain acceptance, Motorola must obtain a license from the FCC to use these frequencies. The basic premise is that a sharing of frequencies will be allowed for RDSS as well as two-way digital voice and data. The use of this L-band will allow the low-powered handsets to communicate within the 420 nautical mile distances with the cells.

Other voice and data services within the domain outlined by Iridium, Inc. will evolve after the basic arrangements are established. Iridium, Inc. recognizes that the requested spectrum will have to increase over the years, but the additional spectrum will not be immediately needed for at least 5 years. Iridium, Inc. has continually sought new partnership and interconnection arrangements with the various carriers and PTTs around the world. The idea is for Iridium, Inc. to license and establish arrangements with these carriers for some fixed time periods, with renewal options, on a non-common carrier basis. Essentially, Iridium, Inc. is trying to deal, not directly with the public, but rather through arrangements with the carriers that will resell the international services. Further, other manufacturers will produce the handsets and add-on devices for sale to the public, the carriers, and the resellers.

The market for the handsets and the add-on devices should prove very lucrative. One could expect that the handsets will operate in a dual-mode capability, in which a user in his or her home cellular or PCS network can use that localized service when available and comparable. However, when the localized service is busy or unavailable, or represents a more expensive option, the user could switch to the IRIDIUM Network to complete a call, message, or data transfer or to locate a device. As other foreign nations recognize the service as viable, the interconnection arrangements will be enacted, thus creating a worldwide "cellular network."

Table 8.4 summarizes the frequency bands requested by Motorola, the immediate requirements for bandwidth, and the projected bandwidth additions that will be needed over time. The requirements may well change as the service gets under way, or as needs dictate. The issue here is the ability to create such a network, not to precisely pinpoint how it will look in 5 years.

The hedge in the last column in Table 8.4 reflects the possible future need for bandwidth, depending on the amount of use and frequency reuse that can be achieved. In the United States alone, the IRIDIUM system will allow for a frequency reuse factor of 5 (5×), whereas on a global basis the frequency reuse patterns will amount to upwards of 180 (180×). This is a step in the right direction, since the results in the frequency band for the RDSS has always been limited to a one-on-one basis. Therefore, the RDSS frequencies in the 2.4835–2.500 GHz ranges with 16.5 MHz allocated for RDSS can be reallocated for other uses. Remember that the operating spectrum for PCS/PCN is in this same limited spectrum and may benefit from an

TABLE 8.4 Bands and Bandwidth for the IRIDIUM System

Band requested	Immediate bandwidth needed	Future additional bandwidth
L-band for set-to-cell communications	5.1 MHz (operating in the 1.610–1.6265 MHz range)	Up to 100 MHz in the L-band for the future
K_a band for gateway feeder links	200 MHz (100 MHz in the 18.8–20.2 GHz range for downlink; and 100 MHz in the 27.5–30.0 GHz range for uplink)	Possibly remain at the 200 MHz, or additional 100–200 MHz for the future
K_a band for intersatellite communications	200 MHz (all in the 22.55 –23.55 GHz range for the intercommunications connection)	Possibly remain at the 200 MHz or additional 100–200 MHz for the future.

expansion of capacities here. The issue remains with the administrators of the spectrum, and falls out of the realm of this book.

Benefits of IRIDIUM Service

Motorola has established a list of benefits from the deployment of the IRIDIUM network, a list that at first glance may look biased toward its services. Further evaluation suggests that these benefits can be derived from any network of this type so long as the authorization is granted. Therefore, generically these will be addressed and kept in the context of any mobile network that has the authority to use the spectrum. Although the benefits obviously lean toward the end user, they can have a significant impact on the provider and the country that allows this connectivity.

Ubiquitous services

With continuous and global coverage, any-to-any connections can finally be achieved as a demand service from any subscriber. As users from any country travel either domestically or abroad, the service travels with them. It will eliminate the need for multiple telephones (i.e., the world cellular standards differ dramatically) special-access arrangements and special numbers that must be dialed to provide the registration and log-on process, as is currently the case on cellular networks. The user should virtually never be out of range from the network. Remote areas that have insufficient demand for services and cannot financially justify the installation of networks and connections will now have the ability to connect anywhere in the world via a single interface—a voice, data, fax, paging, or RDSS service made immediately available.

Spectral efficiency

As already mentioned, the frequency reuse patterns for the bandwidth allocation will be significant: a 5× reuse pattern in the United States and a 180× reuse in the world. No other satellite system has achieved these reuse factors to date. The IRIDIUM System is the first of many to claim such a high degree of transport efficiency, which is sorely needed throughout the world. Since the RDSS portion of the IRIDIUM network is contained in the same spectral arrangement as the other services, the 16.5 MHz of spectrum already allocated to RDSS can be taken back by the regulatory bodies and reallocated for other services. Once again, this is a giant step in the efficient use of the spectrum. It is estimated that to achieve a geosynchronous satellite design providing the same service would require up to 120 MHz of uplink and downlink spectrum.

Public benefits of flexible design

The digital technology that will be deployed with the IRIDIUM network (or some other service, yet to be defined) will allow the total connection for all voice and data services on a 7-day by 24-hour basis in any country allowing the capability. However, flexibility of service is provided. If a country has not authorized access to this network, these frequencies can be shut off while the satellite passes over that country. Further, the height of the system allows for better transmission design. The low-earth orbit overcomes some of the limitations of the higher transport systems, such as the reduced delay in round-trip transmission. Also, since the satellites are low enough, the user set needs a lower-power output device than a mid-earth or geosynchronous orbiting device (MEO or GEO). At greater than 600 miles the system needs much higher output capacities, and at less than 200 miles the orbit requires much more fuel consumption. Therefore, the 420-mile orbit has been selected as the most flexible.

Potential to save lives

How often have the news media published stories of stranded or desperate people who met their demise because they lacked basic life-support systems (water, food, shelter, and medical attention) particularly after an earthquake, flood, or other natural disaster? If only those people had had a means of notifying authorities or rescue parties, their lives may have been spared. A ubiquitous telecommunications service could save lives, or at least increase the odds of people being rescued by turning the confines of an office building, a household, or even a mountaintop into a boundless area where help is only a phone call away.

Vendor capabilities

Motorola has been one of the major players in the production, research, and development of private mobile radio services. Involvement in satellite-based subsystems, under contracts with the military and government agencies, adds to Motorola's depth of experience. Further, the company has been a leader in the development and sales of mobile telephone switching offices (MTSOs) and cellular handsets around the world. These combined capabilities do carry a lot of merit. The value-added services of the IRIDIUM technology are a logical extension of Motorola's combined expertise.

Promotion of international communications

The IRIDIUM network will deliver modern digital transmission services to unserved areas of the world: polar regions, emerging areas that lack a wired communications infrastructure, and maritime regions in which coverage either is spotty or lacks the robustness of a telecommunications service. The FCC and the U.S. government are attempting to use telecommunications as a strategic and economic tool to foster development in these areas. Their goals include:

- Promotion of the free flow of information worldwide
- Development of innovative, efficient, and cost-effective international communications services that meet the needs of users in support of commerce and trade development
- Continuous evolution of a communications service and network that can meet the needs of all nations—and, specifically, developing nations

The above goals can be met with a digital mobile communications network such as proposed in the IRIDIUM system. A wireless network will aid in quick deployment over the more expensive and time-consuming wired technology. The LEO-type services proposed by Motorola have other benefits, many of which are derivatives of the ones listed above. However, these services would be addressed by an individual's specific needs and desires. In addition, the global services mobile (GSM) standard is emerging throughout the international arena as an alternative or a complement to the networks being discussed here.

Proposed Market Penetration with IRIDIUM

Motorola has done its homework in attempting to gain the edge over the competition and to gain approval for the licensing and operation of the IRIDIUM network. It is important to note that the IRIDIUM is not a replacement for existing cellular and emerging PCS/PCN technologies, but rather a complement to them. Therefore, Motorola has been very conservative in projecting how many users may well see a benefit to the services. Initial penetration for the IRIDIUM network is expected to exceed 6 million users by 2001—the fourth year of operational startup. Table 8.5 shows the breakdown by service area. These estimates, prepared by Motorola for the initial license request, will obviously grow over time.

In another slice of the projections for this market, the types of service may vary with user demand. RDSS, for example, are projected to exceed 3 million users. However, all voice and data users would automatically get RDSS as a bonus (part of the price).

Terminal equipment

The terminal equipment for the IRIDIUM network—called the IRIDIUM subscriber unit (ISU)—will consist of a portable handheld unit, a

TABLE 8.5 Potential Users of Iridium by 2001

Service area	Potential users
RDSS tracking	1,550,000
RDSS global paging	1,500,000
Global government communications	
Federal	420,000
State	168,000
Local	588,000
Travel	
International	150,000
Commercial flights	5,000
General aviation	100,000
Business aviation	9,000
Recreational vehicles	500,000
Pleasure boats	420,000
Shipping organizations	46,000
Oil and mineral exploration	20,000
Public telephone services	300,000
Domestic business use in developing countries	300,000
Total	6,076,000

SOURCE: Motorola, Inc.

vehicular mobile unit, a transportable unit and a pager. The basic service offering will include global RDSS and voice and data communications. The lightweight ISU is designed to be used in a manner similar to a cellular phone. Motorola's experience with regular low-cost cellular sets has lent credibility to the venture.

The company has already invested significant funds in the development of the ISU. However, because of the uniqueness of the architecture of the satellite-based service, the costs are not expected to drop as dramatically as did the price of the original cellular sets. Motorola will likely open its architecture to other manufacturers to design and produce compatible devices through a homologation (type acceptance and registration) procedure. Table 8.6 lists the initial prices for the

TABLE 8.6 Preliminary Costs for IRIDIUM Terminal Equipment

Device	Initial cost
Basic handset	\$2000
Pager	\$200–\$300
RDSS-only set	\$200
Data add-on module	\$1000–\$1500

Figure 8.4 The basic handset, or ISU, resembles a cellular phone.

basic telephone set and the ancillary devices. These are preliminary costs for the equipment; the market will likely dictate truer costs as the service is finally deployed. The pattern holds for any new technology: The pioneers sign up for the service at higher costs, but as mass production and competition take hold, prices drop.

The basic telephone set is pictured in Fig. 8.4 as Motorola's conceptual design for the lightweight service access point. As mentioned, the set resembles a regular cellular phone and will be as simple to use. Enhancements will likely reduce the size or change the shape as time goes by. The standard RS232 port built into the side of the phone may well resemble a personal digital assistant. In Fig. 8.5, a docking-type station adapter is added to the set. This is not final in design but would follow the general industry direction for such equipment.

Deployment and spacing of satellites

Clearly, Motorola's concept of multiple orbits on a polar path raised the prospect of midair collisions if the spacing was not correct. Therefore, a good deal of the spacing design had to take into account

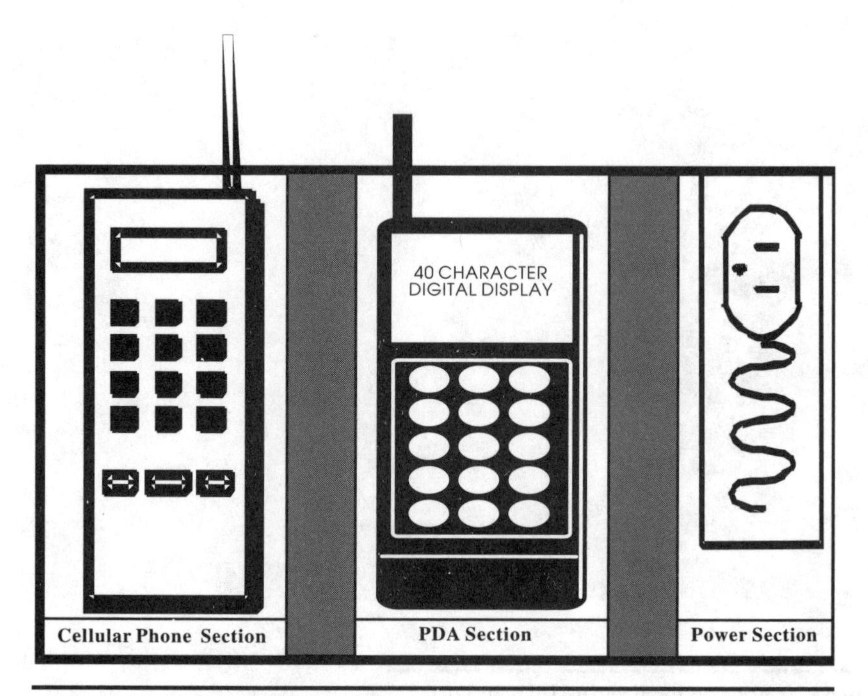

Figure 8.5 A data add-on module may resemble the docking station for Personal Digital Assistants of the future.

the fact that all the satellites would traverse the same end points, where the paths would cross. Essentially, the two poles of the earth would be the midpoints of the orbit. The spacing would have to be carefully planned when all the satellites crossed over the poles. The network originally consisted of 77 satellites; however, a modification dropped the number of craft to 66. This is still a lot of satellites moving around the earth. Motorola took into account six critical criteria in determining the best approach:

1. Satellite coverage over the entire earth's surface must be available at all times. This means that each subscriber, at no matter what location, must always have at least one satellite in view.
2. Some portion of each orbit must be available to allow for low-power outputs to accommodate the recharging of the communications power subsystems. Recharge capability minimizes the size requirements of the craft.
3. The relative spacing of the satellites and the line-of-sight relationships must allow the on-board systems to control cross-system linkage.

4. Costs for the entire constellation of satellites was a primary concern in the selection of the orbit and spacing requirements. Minimizing costs so that compact, low-cost craft could be used was a portion of the decision-making process. Further, the size of the spacecraft would dictate the cost of the launch vehicle, which would add to the deployment expense.
5. The angle of incidence from the end user to the spacecraft, as measured from the horizon to the line-of-sight communications, had to allow for link margins to accommodate the low-powered user device. The slant angle was selected at 10° to meet this criterion.
6. The operational latitude of the spacecraft would affect both costs and design. Satellites operating over 600 nautical miles are subject to greater radiation, which drives the cost up for the systems. Satellites operating under 200 nautical miles require much more fuel and positioning control, again driving up the cost. An optimal position—balancing fuel consumption, control over craft position, and radiational performance—was selected at 420 nautical miles.

The satellites will be spaced 32.7° apart, will travel in the same basic direction, and will move at approximately 16,700 miles per hour from north to south and 900 miles per hour westward over the equator. Given this path, each satellite will circle the earth approximately every 100 minutes. At the equator, a single craft will provide coverage. As the satellites move toward the poles, overlap will occur (more than one craft will be visible), allowing increased levels of coverage above and below the equator. The satellites are expected to have a life cycle of 5 years, unless improvements can be made. The small-craft design allows for the immediate and cost-effective replacement of any out-of-service equipment and for replacement for decaying orbits over time. Each satellite is designed to be independent in its orbit, so that expensive mass change-outs of the craft will not be required.

Basic System Design

A lot has been said about the IRIDIUM system and its features. Now is the time to pull it all together. Since the IRIDIUM network uses much of the wireless communications technology (present and future) described in previous chapters, a summary is in order:

1. *Basic wireless communications services.* Obviously the IRIDIUM network meets the general criteria for a wireless system.
2. *Radio-based systems.* Both microwave and satellite communications are brought together in the IRIDIUM concept.

3. *Light-based systems.* The IRIDIUM network is not designed around light, as are infrared systems. Purely radio systems are in use here.
4. *Cellular communications.* Frequency reuse around the country and the world has become very important. The IRIDIUM network is a spinoff of the cellular communications concepts of lower-powered devices and frequency reuse in small areas.
5. *PCS/PCN.* The go-anywhere, call-anywhere, from-anywhere concept certainly applies to the IRIDIUM network. Global connections will be made to a single user device, no matter where the caller or called party is. Certainly, this is an application for the network to address the user rather than vice versa.
6. *Cellular versus PCS technologies.* A key issue in the comparison of the two technologies is whether the services can complement each other rather than compete. The IRIDIUM concept merely enhances the complementary process. Dual-mode services may be employed so that the user can communicate first with a cellular or PCS set and then, if necessary, switch to an international or global access.

In short, the IRIDIUM system uses proven technology for radio transmission in the operating frequency bands that have already been established. The system building blocks are:

- A space segment, comprised of the constantly moving constellation of 66 satellites in a low-earth orbit
- A gateway segment, comprised of global earth stations that support the interconnection to the public switched telephone network to process calls
- A centralized system control facility
- A launch segment to place the craft in the appropriate orbit
- A subscriber unit to provide the services to the end user.

The space segment

As mentioned earlier, the space segment consists of 66 small satellites operating in the low-earth orbit. All are networked as a switched digital communications system using cellular principles to provide maximum frequency reuse. Figure 8.6 outlines the beam coverages possible with IRIDIUM. Each satellite will use up to 48 separate spot beams to form the cells on the earth's surface. Multiple, relatively small beams will yield satellite antenna gain and reduced RF output. The spatial separation of the beams will increase spectrum

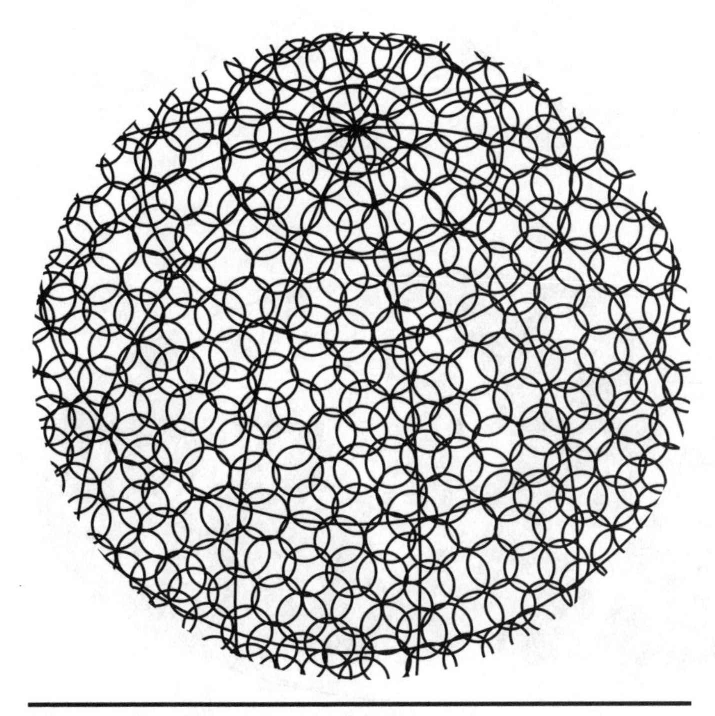

Figure 8.6 Spot beams provide global coverage.

efficiency through combined time, frequency, and space reuse over multiple cells, allowing for simultaneous conversations over a given frequency channel.

As mentioned, the IRIDIUM system is a form of reverse cellular communications. In cellular operations, fixed cell equipment provides service to a large number of mobile users. In the IRIDIUM network, users appear to be stationary while the cells move. Each satellite operates cross links to support the internetworking with its peers. The satellite has forward- and reverse-looking links to the two adjacent satellites in the same orbital plane. Each craft can communicate with the earth-based gateways, either directly or through the cross links of other satellites. The system is designed to operate with up to 250 independent gateways, although in its initial deployment only 5 to 20 gateways will be available. Figure 8.7 presents an overall view of the network communications through the gateways.

Cell patterns. The satellites can project 48 spot beams on the earth, using the L-band frequency assignments. The spot beams form a series of continuous, overlapping patterns. The center spot beam is surround-

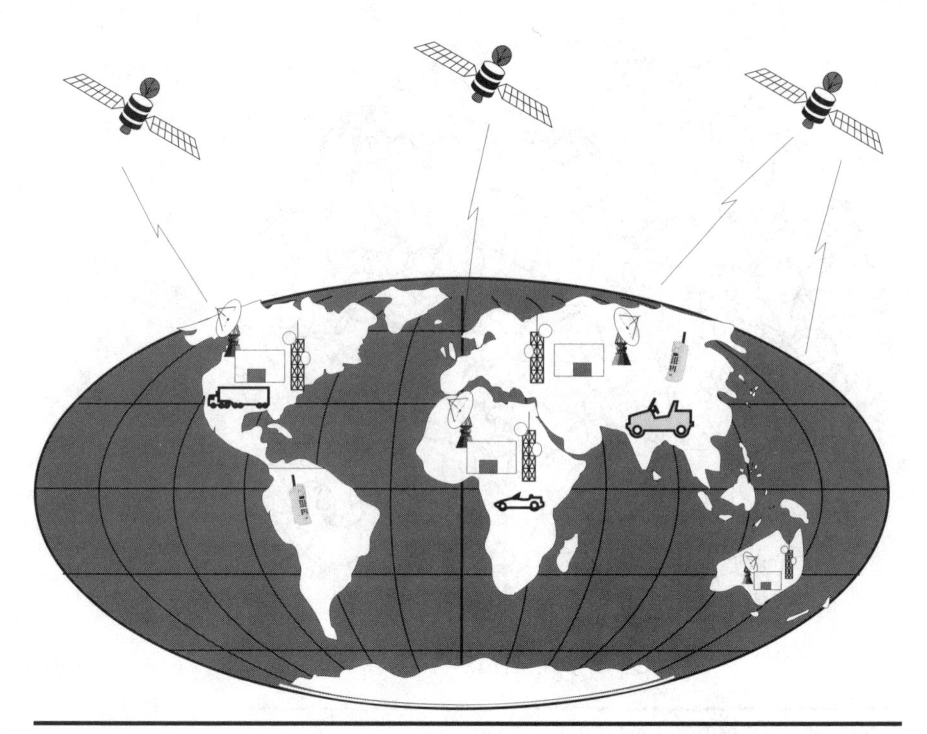

Figure 8.7 The satellites will communicate with the gateways distributed around the world.

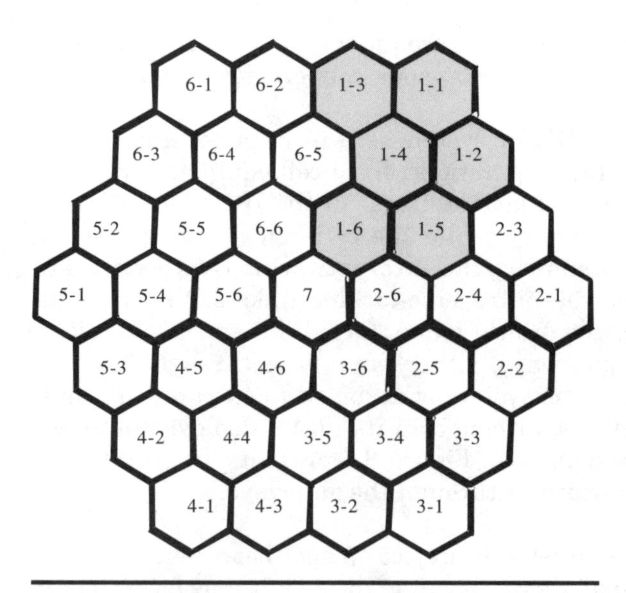

Figure 8.8 The 37-cell pattern formed by the spot beams.

ed by three outer rings of equally sized beams that work outward from the center beam in clusters of 6, 12, and 16. This 48-cell pattern is shown in Fig. 8.8, using the center beam as the starting point. Each of the spot beams is approximately 372 nautical miles in diameter, and the beams combined cover a circular area approximately 2200 nautical miles in diameter for each satellite. A satellite is visible to a subscriber for approximately 9–10 minutes. A 7-cell frequency reuse produces the actual pattern. As shown in Fig. 8.9, the antennas are scanned for communications services on the satellite. The equipment scans in a pattern for service requirements under a time slot arrangement.

TDMA format. Figure 8.10 shows a time division multiple access arrangement. The TDMA time slot is comprised of a 60 millisecond (ms) frame showing the applicable timing sequence in this framed format. For each transmission, the transmit burst time is 1.3 ms. The receive burst time is larger, with a timing sequence of 2.9 ms. The overall time between bursts is 4.2857 ms, with an additional guard band timing sequence of 42.857 μs.

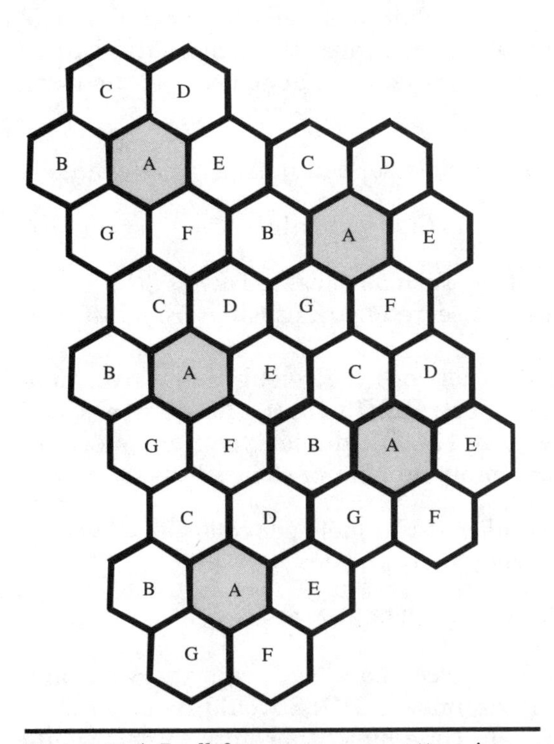

Figure 8.9 A 7-cell frequency reuse pattern is formed for each satellite.

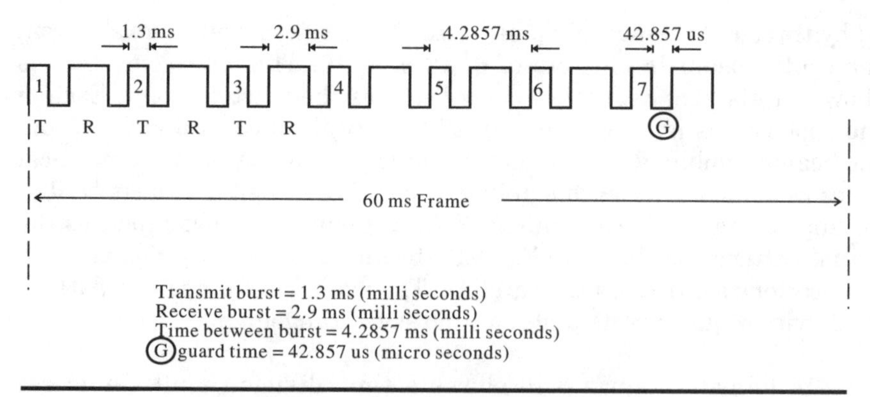

Figure 8.10 The TDMA format in a 60-millisecond frame.

Traffic-carrying capacity. The IRIDIUM network will use time division multiple access *and* frequency division multiple access to produce efficient use of the spectrum. A 14-slot TDMA format allows each cell to be assigned an average of two time slots. The average traffic capacity with a utilization of 10.5 MHz is two times 87 traffic channels, or 174 full duplex voice channels per cell. The IRIDIUM network will have approximately 40 cells over the continental United States at one time. Therefore, for the 48 contiguous states, the math works out as follows:

$$\text{Number of cells} \times \text{Number of channels per cell} = \text{Total voice channels}$$

$$40 \times 174 = 6960 \text{ (approximate)}$$

Other areas of the United States—such as Alaska, Hawaii, the Virgin Islands, and Puerto Rico—will be served by the system from different cells.

The fully deployed IRIDIUM network will yield impressive numbers in a mere 10.5 MHz of spectrum. In total, the numbers will approximate 1628 global cells at 174 full duplex voice channels per cell, yielding a theoretical maximum number of channels as follows:

$$\text{Number of global cells} \times \text{Number of channels per cell} = \text{Total voice channels}$$

$$1628 \times 174 = 283{,}272$$

Again, these are for full duplex voice channels; paging services and RDSS would yield much higher numbers. RDSS would require only a single burst of data in a single time slot. Therefore, a cell would exchange two bursts every 60 ms, or 33.3 bursts per second. Thus, with the exponential capabilities on the RDSS portion of the network, the 40 cells in the United States would yield 1333 (40 × 33.3) observations per second. Taking this to its logical extension, the global numbers would yield close to 200 million observations per hour.

RDSS. The system has been designed to provide RDSS plus voice and data services using all-digital transmission in a time and frequency division multiplexing scheme. In RDSS, an electronic calculation is made of the stationary position of the IRIDIUM subscriber unit relative to a satellite orbit. At present, the position of the ISU can be plotted to within 1 mile. As enhancements of global positioning services come along, the calculation will be improved to within approximately 3 meters. Clearly, such enhancements will add to the strength and power of the service. If global positioning can come to within a 3-meter locator, it will be a demand service that many maritime craft and aircraft subscribers want, particularly when emergency distress services are required.

Modulation techniques. The modulation process and the multiple-access capabilities of the IRIDIUM network are modeled after the latest techniques in terrestrial cellular networks—in particular, the international GSM standard and the U.S. digital cellular techniques. Combining frequency and time division multiple access, the system also uses a data or vocoded voice and digital modulation technique. Each subscriber unit operates in a burst mode using a single carrier. The bursts are controlled to appear at the precise time so they can be properly integrated into the TDMA time frame, which has 14 time slots. Each subscriber unit will be appropriately timed so that its burst is received at the satellite in the proper time slot.

This digital transmission format produces a lower signal-to-noise ratio than could be achieved with an analog transmission. Also, the system uses a differentially encoded, raised cosine, filtered QPSK modulation scheme. All of this is designed to use less spectrum, keep channels closer together without undue interference, and allow for acceptable levels of intermodulation. In effect, the TDMA/FDMA combination with QPSK allows the IRIDIUM network to maximize system and frequency use while minimizing costs as much as possible.

The gateway segment

The gateway segment controls user access to the network and provides interconnection to the public switched telephone network (PSTN). In the conceptual model, multiple distributed gateways around the world will provide the access and control functionality.

Earth terminal. The key to the earth terminal function is the uplink and downlink capabilities with three separate RF front ends supporting continuous service and high up-time availability. One of the RF front ends is used as an active up/downlink with one satellite while a second RF front end is used to establish communications with the next active satellite. The final front end serves in a backup capability in case of equipment failure and provides for diversity against any of the atmospheric conditions (sunspots, equinoxes, etc.) that can plague a typical satellite communications system and cause degradation of service.

One can imagine the need for a highly reliable communications system here, with the any-to-any concept in full regalia. Each RF front end consists of a K_a band antenna, a transmitter, a receiver, a modulator/demodulator, and TDMA buffers. Since the satellites are the moving targets and the gateways are fixed, the antennas follow and track the two satellites nearest to them. Active communications channels are handed off from the current satellite disappearing from view to the next one coming into view. This handoff is designed to be transparent to the user.

Switching equipment. Each of the gateways will house the switching equipment needed to interface between the communications payload in the K_a band and the voice and data channels from the PSTN to handle call setup and teardown as well as call maintenance. The switching systems do the following:

- Transfer common channel signaling information from the PSTN to the RF portion of the IRIDIUM network
- Transfer line and address signaling information from the PSTN to establish circuit switched calls
- Supply in-band tones and announcements to PSTN users calling onto the IRIDIUM network with necessary progress tones and conditions
- Digitally switch pulse code modulation (PCM) digital 64 Kbps standard signals between channel derived from the terminal channels to the PSTN and provide channels to support the necessary in-band signaling capability for call control and progress.

Interconnecting to the PSTN. The IRIDIUM System will provide both voice and data connections through its satellite network, allowing information transfer between two IRIDIUM users, between a PSTN user and an IRIDIUM user, or between an IRIDIUM user and a PSTN user. Voice connections are designed to be fully compatible with applicable ANSI T1 standards for the United States and the Consultative Committee for International Telegraphy and Telephony

(CCITT) G and Q recommendations (international T1/E1 standards) for digital transmission systems using either SS7 or R1 signaling. Likely connections will include domestic local, domestic long distance, and international access to allow for worldwide connections. The data channel specifications will be compatible with the open systems interconnect (OSI) model, a seven-layered architecture developed and recommended by the International Standards Organization (ISO)—that allows for transparent data communications transfer between and among systems. The specifications will also meet the recommendations of CCITT study groups V and X for the standard communications interfaces and electrical and mechanical interfaces for data communications equipment.

The system control segment

The system control segment (SCS) of the network provides the necessary control over the entire satellite constellation. The SCS manages and controls all the system elements, providing both short-term and long-term service. The functions of the SCS are divided into active control of the satellites and/or control over the communications assets within the satellites. The SCS monitors the overall network control sequence, balancing traffic loads and managing peak conditions as necessary. The system also handles equipment failovers to maintain the overall efficiency and integrity of the network. The SCS will be a redundant operation, replicated at two separate sites to prevent any catastrophic failure or loss from a single event. Only three terminals will be required to see every single satellite in every orbit. The last stage of the SCS is the long-term planning and analysis for the IRIDIUM network, which will be housed in Arizona.

The launch segment

The launch segment is responsible for positioning each of the satellites in the correct orbit (parking place). This includes the total picture of the launch vehicle and the satellite, and the control of the launch operations. The satellites may be launched singly or in clusters, depending on the vehicle chosen and the overall economics of the scenario. Once the launch is completed and the satellite is delivered into the appropriate position, the SCS takes over control.

The subscriber unit

The subscriber unit segment of the network will consist of communications between the "handset" and the satellite over a full duplex FDMA channel in TDMA bursts of QPSK-modulated digital data.

Motorola will provide the necessary equipment to digitally encode the voice and data transmission to and from the handset. User 2400 baud data and 4800 bps digital voice data are encoded and interleaved to provide the necessary integrity of the transmission. The uplink burst timing is synchronized to the downlink bursts, allowing for the necessary adjustments of location and timing to fit the transmitted signal into the appropriate TDMA time slot.

Timetable for Delivering the IRIDIUM Network

Obviously, the network deployment of 66 satellites in a polar orbit, coupled with all the orchestrated control and funding, will take a while to deliver. At the time of filing, Motorola stated that the first satellite would be launched in 1996. In 1993 Motorola began testing the communications handoff capabilities by using high-speed aircraft (planes) which approximated the same speed as the satellites would traverse the earth's surface. Plans are continuing. However, the first launch is only the beginning piece. Table 8.7 summarizes the overall plan to date. Some slippage may occur in the plan, but Motorola has announced its intention to stay as close to the targets as possible. In part, this adherence to a stringent schedule may be the direct result of competition on Iridium, Inc.'s heels. Any slippage in the delivery could cost significant dollars and also allow a competitor to get there first and grab the market share. Thus the targets are as firm as one could imagine.

Competitors to the IRIDIUM System

As mentioned earlier in this chapter, several other suppliers have petitioned the FCC and the WARC for rights and licenses to launch satellite constellations in low-earth orbits, mid-earth orbits, or geosynchronous orbits to provide any-to-any communications—the personal communications services of the future. Each entrant stands to gain an edge if it can launch and operate first. These competitors, including

TABLE 8.7 Target Dates for IRIDIUM System Milestones

Milestone	Target date
Construction of satellites begins	1992
First satellite completed	1996
First satellite launched	1996
All satellites constructed	1996–1998
All satellite launches completed	1998
Initial operations begin	1998
Service begins for customers	1998

TABLE 8.8 Vendors Competing in the Satellite Arena

Company	FCC license	Proposed launch date	Cost per minute for use	Technology
AMSC	Yes	1994	$1.50	FDMA
Constellation	Pending	—	$1.50–$2.00	CDMA
Ellipsat	Yes	1995	$ .50	CDMA
Loral-Qualcomm	Pending	1997	$ .65*	CDMA
TRW	Yes	1998	—	CDMA
Iridium, Inc.	Experimental	1994	$2.10	FDMA/TDMA

*The Loral-Qualcomm Service is $0.65/min plus the long distance call. This network does not use intersatellite handoff so the call must be brought down to a gateway, then connected to the long distance telephone network (international calls average $1.25–$3.00 per minute).

some of the big U.S. names in the business, are all vying for the licenses to use the L-band frequencies to provide global communications.

Table 8.8 lists some of the most active players, along with the service they intend to provide. The table is intended, not to be all-inclusive, but to give the reader a comparison of the various approaches. With their diverse service offerings, the chief vendors are competing either to cover remote or rural areas or to complement the PCS services for global communications. Here is where they stand in the total picture.

American Mobile Satellite Corporation, Washington, DC

AMSC has proposed a GEO arrangement to provide global service in fixed slots for round-the-clock coverage. Under the GEO concept, only three satellites will be required to provide the proposed coverage. However, the cost of the launch is far greater than that of the LEO and MEO arrangements. AMSC has already been granted a full license by the FCC to begin its operation and launch proceedings. AMSC plans to offer its service in the United States only (including Alaska, Hawaii, and Puerto Rico). The company stands to be the first and closest to providing true domestic PCS services—if it can get the funding and launch in order quickly. As noted in Chap. 2, GEO provides significant amounts of bandwidth for voice, data, video, image, and CATV services. The operating life is now approaching 10 years; therefore, the GEO service will require far less replacement than the other orbits, which tend to decay quicker.

TRW, Redondo, CA

TRW has proposed a MEO arrangement under the network name of Odyssey. The plan calls for launching some 12 satellites in a mid-earth polar orbit (1100 nautical miles above the earth) and for implementing a code division multiple access (CDMA) technology to maximize frequency reuse. The deployment of 12 satellites is a more efficient spectrum use, according to TRW. The Odyssey satellites will provide global coverage for longer periods of time than the IRIDIUM network (estimated at 10 minutes), although the height of the satellites more closely resembles a LEO orbit than a GEO orbit. At a height of 1100 nautical miles, the footprint of the satellite will be larger than that of a LEO, so coverage will require fewer satellites. However, the power requirements, the RF output, and the size of the vehicle are all larger than in the IRIDIUM system.

Loral-Qualcomm, Palo Alto, CA

Loral-Qualcomm's approach is to use a big LEO arrangement similar to Iridium, but with fewer craft (approximately 48 satellites). The network handle is Globalstar systems. Qualcomm expects to use its expertise in the specialized mobile radio business and a code division multiple access (CDMA) arrangement that the company pioneered. This service is expected to begin around 1997. Loral-Qualcomm will draw from its experience in the business by deploying its CDMA technology and will attempt to reuse the frequency spectrum as much as possible. Service offerings will include fleet tracking, locator services, and ultimately, by logical extension, voice and data communications.

The uniqueness of the Loral-Qualcomm system is that it is designed to fully complement the cellular and PCS industries by providing single-service coverage, from a single phone to a single number, through an intelligent network. Unlike Motorola and the other pioneers, Loral-Qualcomm does not plan to use a dual-mode operating set. Rather, a single set will interface to any network service without having the user make a decision. Also, the Loral-Qualcomm network plans to offer the same cellular or PCS service for around the same cost per minute ($0.65) as the other networks for airtime plus the cost of long distance terrestrial service as noted.

Constellation Communications, Fairfax, VA

Constellation is a consortium-based service comprised of various organizations, including Pacific Communications Services, Defense Systems, and Microsat Launch Corp. These players have their own technology and a base of applications and equipment at the ready. The system, called Aires, is a LEO arrangement using 48 satellites and CDMA technology to maximize frequency reuse wherever possible. The funding issues remain unfinished, with an initial cost slated at approximately $200 million. Very little has been stated in terms of the overall plan or service niche, as well as the company's licensing status. About all that is known is that Constellation intends to enter the market. The rest is something of a mystery.

Ellipsat, Washington, DC

Unlike the other licensing candidates in the market, Ellipsat does not have a technology of its own. The system will be comprised of satellites and network engineering created by others (Fairchild Industries and Israel Aircraft Industries). Launch is planned for late 1995, with implementation of the network set for the middle or end of 1996. These are fairly aggressive plans at best. The Ellipsat network, known as Ellipso, will use approximately 24 satellites and CDMA technology. The estimated cost is $450 million to $500 million. The network strategy is to penetrate the market, not with technologies, but with service offerings that fill the cellular gap. Ellipsat plans to concentrate on the land-based coverages for the cellular carriers rather than attacking air and maritime services. Its goal is to complement the cellular market. Therefore, Ellipso is attempting to keep the cost for service at rock bottom. Initial estimates of $.45–$.50 per minute are competitive with the cost of cellular communications.

Little LEOs

Other niche marketeers filing to license their services are primarily in the locator systems. Services include two-way messaging and spot filling for the cellular carriers to give total connectivity through the interconnection of the service. Many of these providers are looking to use either VHF or UHF licensing, since the service does not require high bandwidth for voice or data communications. These are known as little LEOs. Two of note are Orbcom (a division of Orbital Sciences) and Starsys, which will use some leased line facilities from Inmarsat, a current player in the GEO arena. The primary services that will be offered include:

- Asset recovery (stolen property)
- Locator systems (fleet tracking for mobile operators)
- Emergency preparedness (disaster recovery messaging)
- Remote monitoring (process control, security, etc.)
- Hazardous-material tracking and handling

TABLE 8.9 Proposed Number of Craft in Satellite-Based Systems

Company	Number of satellites proposed
AMSC	3 GEO
TRW (Odyssey)	12 LEO
Loral-Qualcomm (Globalstar)	48 LEO
Ellipsat (Ellipso)	24 LEO
Iridium, Inc. (IRIDIUM)	66 LEO
Orbcomm	3 little LEO
Starsys	3 or more little LEO/GEO
Constellation (Aires)	48 LEO
Others (VITA and CLS service; Argos not mentioned)	N/A
Total new satellites	201 LEO plus 6 GEO at a minimum

The industry will be alive with the various offerings over the next few years. Think of all the satellite systems that have just been discussed! The numbers are getting quite large in the space segment, as shown in Table 8.9.

Chapter

9

Wireless LAN Applications

Overview: Future Wireless Applications

The preceding chapters dealt with the emergence of the wireless-based system around the world. These technical and not so technical discussions were geared to provide an understanding of radio- and light-based systems. We turn now to the application of the technologies in everyday life, primarily in the business and government sectors. Recall that any technology is only as good as the service that it provides. Too many technical solutions have been manufactured and marketed without a real sense of the problem that was being addressed. Therefore, the remainder of this book will attempt to make the different applications fit a little better.

This chapter examines the complementary services achieved from wireless connectivity. Chapters 10 and 11 address the competitive nature of the offerings. The presentation is in no way meant to suggest that the technologies must compete with each other. Rather, the end user must seek out and apply specific solutions to business-related problems or opportunities. Regardless of the actual methodology used, there will always be other opportunities to research. The market is too dynamic merely to look at a single solution. The up-and-coming derivatives of each service have paved the way for a myriad of options. It should, therefore, be incumbent upon telecommunications professionals, administrative managers, designers, and controllers to look at every possible selection before committing to a single solution.

Further, the ability to mix and match services and technologies still reigns. No one service may totally displace another when the systems can coexist harmoniously in an ever-demanding and increasingly flexible environment. The use of wireless solutions may fit a niche, rather than be used as a single replacement technology. The vendors responsible for the manufacture and sales of these systems and technologies are somewhat guided by the need and desire to sell the services they offer. They will have us believe that theirs is the only solution to a problem, while we as novices tend to let them be our guides. Too many organizations have been led down the path of misconception and limited options because of lack of total understanding.

Organizations also get caught up in the "latest and greatest" attitude—a mindset that can cloud their vision. This is not a criticism of the industry, but a warning that all options must be thoroughly researched. Many of the systems that may fit a niche are de factor standards, created by the purveyors of the system. Until the industry gets serious about standardization and interoperability, the risk always looms that the solutions we select today will be the stranglehold on our adaptability in the future. Be aware! There are some very fine applications-driven solutions, and there are many opportunities to put wireless communications to work in an organization. As long as the systems are understood, there should be no problem that cannot be addressed and overcome in the future. So the applications and the services combined should drive the decision-making process.

Financially the need to justify the systems will always be a requirement. Therefore, keep in mind that the newer systems will be more expensive initially than when they are proliferated around the country or the world. The discussion of PCS clearly indicated that in time the costs of personal communications may well drop to a reasonable $10.00 per month and $.10 per minute of use. However, as these services are introduced, they may be equally as expensive as, or more expensive than, cellular communications. The point, again, is that pricing issues must never be overlooked. It is the pioneer who will be the innovator and the experimenter until the "bugs" are worked out of any emerging service. Then the masses will adopt a standard offering at a far more reasonable price.

The futuristic discussions of wireless technologies and services, provided in this and the remaining chapters, are in no way intended to excite the reader into jumping into an unproven application, or becoming a pioneer, when the organizational approach is a "wait and see" attitude. The logical framework described is merely *what might be,* not what will be. It is still up to the vendor community and the standards bodies to determine in which direction the market will go.

No one can say for certain just what the opportunities or the standards of the future may be in wireless systems. For every new service, a counter will likely be created with a wired technique. For every new technique, a concern group may emerge to resist its deployment. An example is the cellular "scare," in which claims that the radio frequency and electromagnetic field supposedly caused cancer in users threw the industry into fearful and reactionary mode. This, of course, will be a topic of discussion and debate for years to come, as the claims of health risk continue to emerge. Such concerns are obstacles to overcome within the workplace or the total environment. Initial dollar investments and longer-term commitments may be required to change out or migrate to wireless services. These will require a certain period of time to pay for themselves. Therefore, the appropriate period of time implies that the system or service is stable, will not have to be yanked out or mass-replaced, and will be available with minimum ramp-up time. If these goals cannot be achieved, a serious reevaluation of the wireless technology is in order.

Table 9.1 summarizes possible concerns and applications as a means of justifying a pilot or wide-scale wireless service. This is not the only approach to the service, but it does meet some of the general criteria for the evaluation process and the justification stages. These criteria should be either replaced or supplemented by organization-

TABLE 9.1 Issues and Concerns in Analyzing New Technologies

Issue or concern	Answer	Points
1. Is this technology capable of being handled by existing services, such as wired systems?	☐ Y ☐ N	
2. Is this technology new in your organization?	☐ Y ☐ N	
3. Are there other options that can be researched?	☐ Y ☐ N	
4. Are other companies using the service?	☐ Y ☐ N	
5. If others are using the service, is their response to the service and installation favorable?	☐ Y ☐ N	
6. Do industry standards for this offering exist?	☐ Y ☐ N	
7. If standards do not exist, are other options or de facto standards being used by other vendors?	☐ Y ☐ N	
8. Are there any legal issues brewing about the implementation of this technology or the application being considered?	☐ Y ☐ N	
9. Does this technical solution compare favorably with others available?	☐ Y ☐ N	
10. How else can the service be handled?	☐ Y ☐ N	
Is this option the preferred solution?	☐ Y ☐ N	
11. Do you understand the physical and electrical requirements of this service?	☐ Y ☐ N	
12. Does the vendor display a full understanding of the technology as it applies to your application?	☐ Y ☐ N	
13. Is this the only available option to solving your needs?	☐ Y ☐ N	
14. If this is the only solution, is it more expensive to install and operate than the current system?	☐ Y ☐ N ☐ Y ☐ N	

TABLE 9.2 Comparison of Wireless Options

Application	Option	Technology
Local area networks	Wireless or wired	Radio or light, coaxial (baseband or broadband), twisted pair (UTP/STP), or fiber
PBX services	Wireless or wired	Radio, twisted pair, or fiber
Mobile telephony	Wireless or pay phones	Cellular or PCS
Data transmission	Wireless or wired	Cellular, packet, or PCS
Facsimile	Wireless	Cellular, packet, or PCS

specific issues, as the user deems fit. A point system should be used with Table 9.1 to evaluate whether or not a technology should be pursued. If, for example, a 10-point system is applied (10 points for every yes answer), any raw score greater than some arbitrary number (say, 100–120 points) may point to a good decision, whereas any score below this arbitrary number may suggest additional research.

Table 9.2 lists some of the options that are coming downstream in the wireless industry. This is not an all-inclusive list but an attempt to get the reader into the right frame of mind before applications are decided upon.

Wired Versus Wireless LANs

Why should a user even consider a wireless LAN? Obviously, this is a question that each individual must answer before taking the next step. Until the recent past, wired LANs were the only option available to provide the necessary connectivity between and among workers in a dynamic environment. The tethering to expensive house cabling systems was tedious and in some cases a LAN administrator's nightmare. Many of the existing wires inside a building were used to connect devices together, with no regard for the need for accurate record keeping or the quality of the cabling systems. But as the need to be flexible and efficient changed within the office domain, the existing wired systems were strained to the breaking point. A wireless LAN may be an expedient to get a user base up and running within a certain period of time—a base that could not be physically provided using a wired system. Further, various social or legal constraints—such as the aesthetics of the office area or the legal requirements for certain types of cable—may dictate the use of a wireless platform. Financially wireless service may not be the least expensive solution,

TABLE 9.3 Comparison of Wired and Wireless LANs

Wired system	Wireless system
Money is an object, and the costs to install wires is relatively low.	Money is no object, and new technologies are always being considered or trialed.
Location is fixed; users will never move.	Location is temporary only, or users will constantly be moving, added, or changed around.
Conduits are empty and cable is easily pulled.	Conduits are stuffed. In order to provide LAN connections, new conduits will be required or existing may have to be unplugged. However, if an attempt is made to clean out existing conduits, the risk of disruption or pulling out wrong cables is high.
No legal requirements or constraints exist in the installation of wires.	Plenum space requires special cables or conduits that are very expensive to install or that cannot be pulled in for lack of space. Asbestos in the ceiling or crawl spaces will require special abatement before a cable can be run into an area.
Access between floors and closets is readily available.	Floors will need to be cored. The process is expensive, may not be easily approved by the building owner, and may produce fire risks. Further, the access to the systems may not be readily available.
Aesthetics are not important. Walls and power poles are accessible and will not pose any problems with the look of the office.	Aesthetically the architects left no place to access cable trays or conduits or any other point along the way. The environment does not lend itself to power poles or raceways that would detract from the office decor.

especially if none of the workstations is ever going to move. If they are scheduled to move, the system can become more attractive than a wired cable system. Table 9.3 lists possible choices to go to a wireless LAN. This is a representative sampling of the considerations that may be accounted for. The list is a straight analysis, not a point-by-point comparison of the wired and wireless services. Table 9.4 lists some potential strengths and weaknesses of wireless LANs. Once again, this is a comparative look only.

Wireless (untethered) LANs will extend the existing wired arrangements, providing unparalleled convenience and mobility for the future. Critics have been quick to argue that the wireless solutions for LANs are too expensive and slow compared with those of wired LANs. Although this is true at a superficial level, things are changing

TABLE 9.4 Strengths and Weaknesses of Wireless LANs

Strengths	Weaknesses
Creates a more mobile environment for the PC and LAN user	Will lower performance below that of a hard-wired system, because of the limitations of bandwidth
Avoids the higher costs of moves, adds, changes, and constant rewiring	May be subject to interference or distance limitations, depending on the technique used
May lower the LAN maintenance and ongoing costs, since there are no wires that require reconfiguration	Requires more expensive and proprietary hardware, depending on the system used

quickly. The wireless vendors have been reducing the costs for their equipment and improving the speed and performance of the systems. The wireless LAN will actually encompass three different approaches for connectivity solutions.

The first solution is access to the wide area networks (WANs) and metropolitan area networks (MANs). In the wider area, the network transmission systems will use the cellular arrangements and the wired long-distance networks as entrance and egress facilities dictate. The packetization of the data will be important to meet the immediate demands of the user community, while the form and format of the data will be important to prevent excessive overhead and consequent latency in transport. The second solution is localized communications services (true LANs) for the added convenience of connections between floors and desktops in a very dynamic environment. Flexibility to provide the quick connects for moves, adds, and changes offers the organization a significant improvement over the basic wired system.

The third solution is the flexible mobile LAN arrangement. As users become mobile within the confines of their own office building, floor, or department, the need to communicate data to and from the client/server devices will be more predominant. This nomadic form of connectivity is emerging in all walks of life and business communities of interest. As the workforce becomes more mobile, the need to provide untethered connectivity increases exponentially.

Buffering of the data is another important development that has enabled the user to move about freely with access to data on demand. In the basic form of the wireless LAN, the user may also have a wireless adapter and a standard network operating system so that, in moving from office to office, or floor to floor, the user is guaranteed access to customer information, working papers, and other forms of data. Some form of access point—whether it is called a telepoint, a control station, or a control module—is necessary to allow the interconnection to the wired LAN. This is a system-level issue that must be addressed by any of the current producers as well as by new entrants into the marketplace.

Two different technical solutions exist with the optional wireless local area networks. These two solutions are radio and light technologies. In the radio technology, two solutions exist again: the licensed microwave radio frequency range (18–23 GHz) or the unlicensed radio frequency range. In the unlicensed radio frequency range (902–928 MHz; 2.4 and 5.7 GHz), there are two additional options. The first is to use a spread-spectrum technique, while the second is to use a spread-spectrum technique with a coded-chip sequence. In the light technology, two options exist using infrared light as the medium to transmit the information. The first is a point-to-point service, while the second is a diffused-light arrangement. Both of these operate in the infrared (100 terahertz [THz]) frequency spectrum and are unlicensed.

Regardless of the technological approach that the vendor takes, there are some basic rules that must be applied before any acceptance will be achieved. The vendor must be prepared to deal with the following crucial points before a user would even consider a LAN system:

1. *Is the system simple to install and use?* If the system is too complex to use or adds to the depth and complexity of an already complex LAN, the user may balk at the introduction of the wireless technique. Regardless of the gains to be achieved, if it is not a manageable solution, then there is no place in the business community for it. Beyond the simplicity of use, the system must be somewhat seamless as users roam from place to place within the LAN environment.

2. *Is the connection secure?* Before putting any data on a LAN, the user will be concerned with the integrity and security of the data. The security issue will always be paramount when dealing with a radio- or light-based system. Because of the inherent broadcast nature of a wire-based LAN, the same characteristics are applied to the wireless LAN. More cases of fraud, theft, and alteration of data on a LAN are surfacing. The ability to jam or corrupt data is also becoming a hot button. More mission-critical and sensitive data are running along the wires of a LAN—information whose loss or alteration in the medium could have catastrophic consequences. Node authentication will be a requirement; so too will scrambling or encryption of the data and other methods to prevent unauthorized access to the network or to the actual user data that reside on the LAN.

3. *Is the system cost-efficient?* Cost may play a lesser role in the initial questions from a LAN administrator or manager, since the

technical solutions must be solved first. However, sooner or later the issue of money and efficiency is going to surface. Using a good or service as long as there is reason to do so is a business proposition. Finally, the technology must pay back in a reasonable amount of time—typically, 3 years. Payback is achieved through the avoidance of moves or rewiring, the prevention of the need to reevaluate new technologies (and change accordingly), and the reduction of day-to-day costs. Thus, approval from management is more likely.

Vendors have taken various approaches to offering wireless LAN connectivity. For example, the offerings cover each of the two technologies (radio and light) and each of the internal options: licensed or unlicensed, spread-spectrum or coded-chip sequence, and point to point or point to multipoint. In each of the cases, the bandwidth is slightly different, the topologies are different, and the costs are different. Merely evaluating these simple options for connecting LANs is a full-time project. Further, as each of the LAN topologies and technologies are agreed upon, the option to let the wireless LAN stand alone or to interconnect it to a wired LAN backbone is still a consideration. The options are as complex as the choices that must be made. Finally, as standards emerge, these options may change in structure or in deployment. The following discussions describe the range of these options and the various vendors that are approaching the new service.

Fixed-Frequency Licensed Microwave Wireless LANs

Motorola's wireless LAN offering, introduced under the trademark of ALTAIR uses the fixed-frequency microwave range (18 GHz) transporting Ethernet* with a CSMA/CA access technique. Motorola's idea was to provide the highest possible bandwidth in a controlled radio frequency. As such, Motorola holds the licenses in the frequency ranges offered on ALTAIR. Motorola authorizes the use of the license in a specific location and environment. To use the service, Motorola will configure a system that operates within a building or other type of location, and grants permission to use the frequencies. To prevent any possible overlap or interference in this frequency range, Motorola holds the license information in a database.

*Ethernet is a registered trademark of Intel, Xerox, and Digital Equipment corporations. In this sense, the Ethernet reference is to a standard 10 Mbps LAN operating in an IEEE 802.3 topology. Whereas traditional Ethernet uses carrier sense multiple access with collision detection (CSMA/CD), AT&T and all the other wireless LAN vendors use carrier sense multiple access with collision avoidance (CSMA/CA).

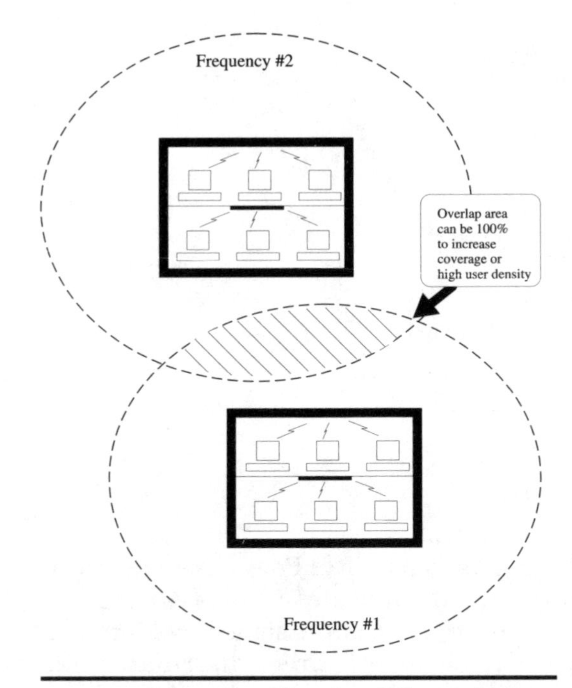

Figure 9.1 Motorola will assign different frequencies if multiple LANs overlap.

Fig. 9.1 illustrates this concept with the coverage area. The arrangement in no way implies that only one sublicensee can have a system in a certain radius. It means only that Motorola will be careful not to issue two separate licenses in close proximity with the same operating frequencies. Thus, interference should be avoided. In the event an authorized user has the need to relocate, Motorola will survey the new location and ensure that no other user is licensed in the area. If, however, another ALTAIR user is in proximity to the new location, the user may change the frequencies in order to prevent any interference between the two users. The setup leaves all the operating and licensing requirements right in the vendor's hands, and relieves the end user of the need to resolve issues of coordination and conflict.

System components

The ALTAIR system has two major components: (1) a Control Module that can coordinate multiple User Modules and (2) a user module that can control up to eight users. In Fig. 9.2 the Control Module is mounted high within the user's space. Typically this device is mounted at

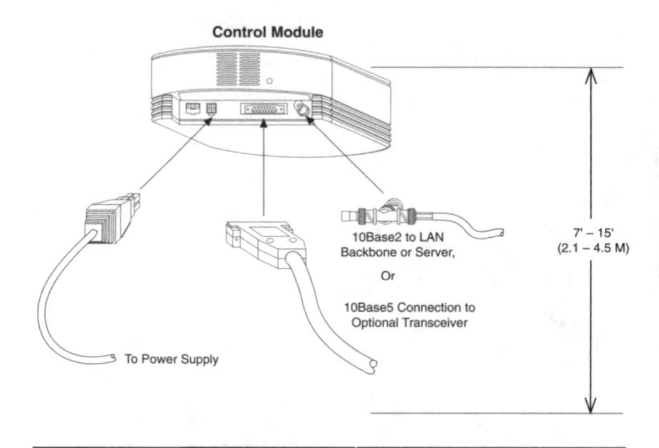

Figure 9.2 The ALTAIR control module is mounted at ceiling height to get the best coverage possible. (*Courtesy of Motorola.*)

ceiling height to get the best signal capability. Power is required at the Control Module. A Control Module is a transceiver (transmitter/receiver) operating on frequencies assigned by Motorola. Whether a value-added reseller (VAR) or direct distributor has installed the system, it is still under the license control of Motorola.

Fig. 9.3 shows the user module. This module is also a transceiver that uses the radio frequencies to transmit or receive the information to the control module. The system is built around a standard Ethernet network interface card (NIC) installed in a PC or other system, with either a 10 base 2 (thin-wire Ethernet) adapter using a

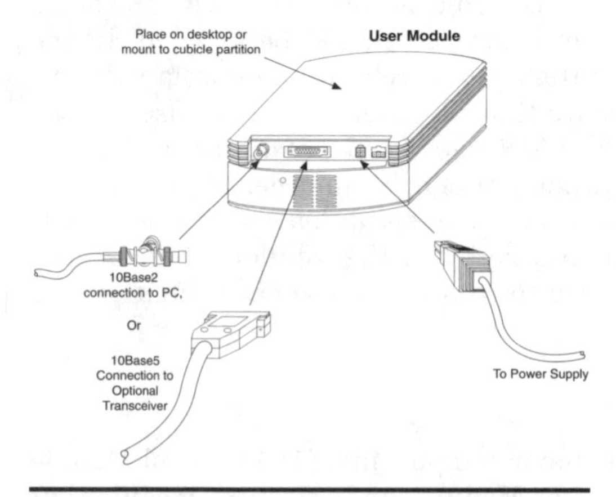

Figure 9.3 The user module is a transceiver that interfaces to the workstation. (*Courtesy of Motorola.*)

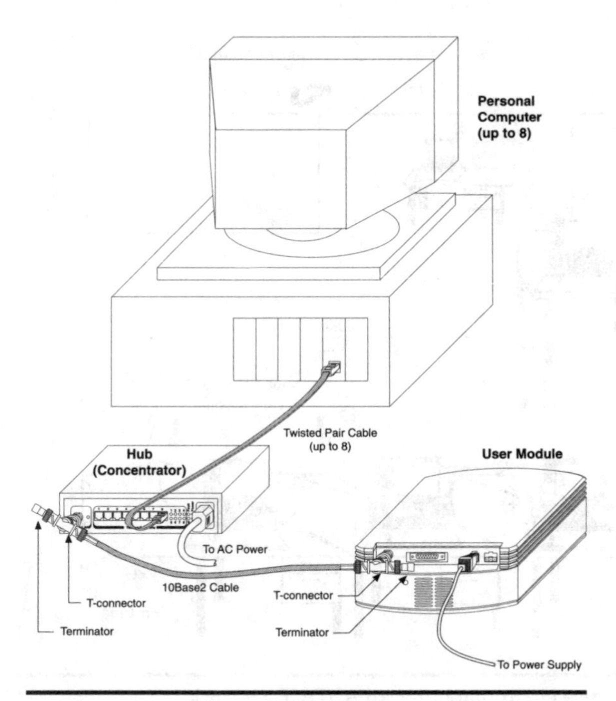

Figure 9.4 The user module acts as an 8-port hub that can support BNC or RJ-45 jacks. (*Courtesy of Motorola.*)

standards-based BNC connector or a 10 base T (10 Mbps on twisted pair) standards-based connector using an RJ-45 jack. Both of these connections are shown in Fig. 9.4. Since a standard interface card is used, the network is independent of protocol and operating systems.

The original ALTAIR user module supported an average of 6 interconnections when the ALTAIR LAN was fully configured. Also, up to 32 user devices were supported by the Control Module. Newer versions, called ALTAIR Plus II, support up to 50 users on a single control module. The user module can plug a single device or can interface to an 8-port hub, allowing greater connectivity. The throughput of the standard Ethernet is set at 10 Mbps. Thus, even though a standard Ethernet NIC is being used, the ALTAIR system initially supported only 3.3 Mbps. The newer ALTAIR Plus II systems now support up to 5.7 Mbps.

Motorola now has the capability of joining two Ethernets together—an installation using their VistaPoint product—at distances of up to 3940 feet. A typical office environment is shown in Fig. 9.5 with a single control module and multiple user modules arranged in and around a partitioned office space. This arrangement is fairly easy to

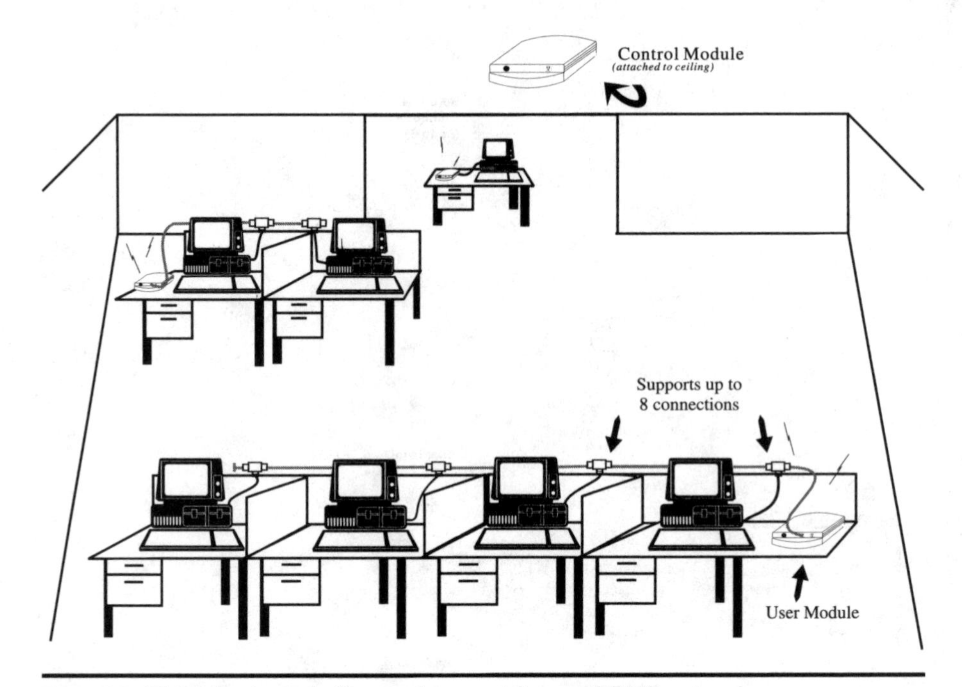

Figure 9.5 Using the typical office environment for an ALTAIR network.

set up and can provide very quick connectivity. The total throughput on this network may not be an issue, since the average usage on the wired Ethernets will likely be less than 40 percent, or 4 Mbps.

It is also possible to hook the control module to a wired backbone service to allow for connectivity to the rest of the organization. This setup is especially attractive when a small departmental LAN is required to move or to interconnect to another organizational LAN. Further, in many of the office environments in which the architects or office planners create a landscaped open plan, access to a cabling system may not exist. The problem can be expensive to overcome if coring of floors is required, conduits are run, or a general power pole arrangement is used. Further, the look of the office could be destroyed if poles are installed everywhere. The obvious solution is to incorporate a wireless connection to the desktop for access to the LAN. The control-to-user-module concept is fairly straightforward and clean in its installation, and fits the bill nicely.

Steps in installation

Installing an ALTAIR system is also straightforward. Since the system uses existing standard network interface cards, no special connec-

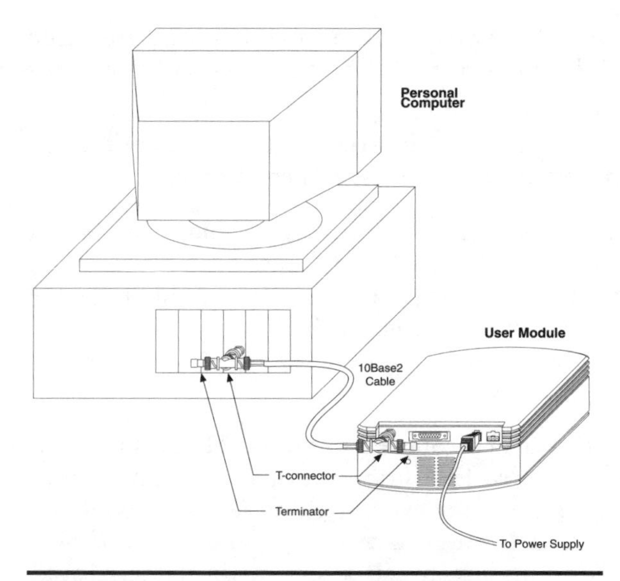

Figure 9.6 The ALTAIR is not totally wireless, the BNC or RJ-45 is a wired connection to the user module. (*Courtesy of Motorola.*)

tivity is required. As may be seen from Fig. 9.6, the ALTAIR is not totally wireless. The standard BNC connector, or an RJ-45 jack, is used to connect to the user module. It is from user module to control module, or from user module to user module, that the medium is wireless. Therefore, no special installation instructions are needed. The user merely opens the system, provides user module and control module connections, and powers them up. The addressing is then automatically logged into the network database and away the system goes.

The following step-by-step installation procedure is achieved by introducing a simple interface to the network:

1. Open the PC, workstation, or other device to be used.
2. Install standard NIC card in the appropriate slot.
3. Close the PC or other device.
4. Attach cable (BNC or UTP) from the back of NIC to the transceiver (user module).
5. Plug in the devices (PC, user module, and control module) and power on.
6. Download drivers for the NIC card to the PC, workstation, or other device. Downloading is for the actual NIC driver and the PC; it is not a requirement for the ALTAIR.

7. Establish communications between the user device and the user module. Step from the user module to the user device to the control module.
8. Make sure addresses of NICs are entered into the network database to provide the necessary addressing and connectivity.*

Even though the ALTAIR system is relatively simple to install, a solid-light indicator is provided as a backup to let the user know that the connection is made. In the event of a failure, the light indicator will not signal. Therefore, quick reactions are facilitated.

Security issues

Securing an ALTAIR network is a concern to many. Motorola uses a scrambling technique to secure the data from eavesdropping and from unauthorized interception. However, nothing is done about the risk of alteration of the data through network jamming or interference, since these interruptions are much more easily detected and traced down.

In the ALTAIR system, the license to operate the frequency brings with it the responsibility of installing an interfering station to take corrective action without debate. Anyone corrupting the data must cease and desist under penalty of law. As the actual licensee, Motorola has the in-house expertise to handle these issues and detect jamming or interference quickly. Further, it has the knowledge of the rules of frequency licensing and use. Thus, Motorola will be responsible to solve this type of problem if the need should ever arise.

Spread-Spectrum Radio-Based Services

Spread-spectrum radio-based systems utilize two different approaches. The first is a spread-spectrum-only sequence that uses the same techniques discussed in previous chapters. Figure 9.7 compares the fixed-frequency and the spread-spectrum networks. Spread-spectrum LAN technology utilizes the 902–928 MHz frequency band—an unlicensed industrial, scientific, and medical (ISM) band that allows manufacturers to supply product with very limited constraints. (Newer products are emerging that use the 2.4 GHz band.) The only major limitations are as follows:

1. The system is restricted to 100 milliwatt output.

*The ALTAIR modules are self-learning and will automatically update the list of devices as they transmit information. The controllers on the wired backbone must be updated to recognize the new addresses.

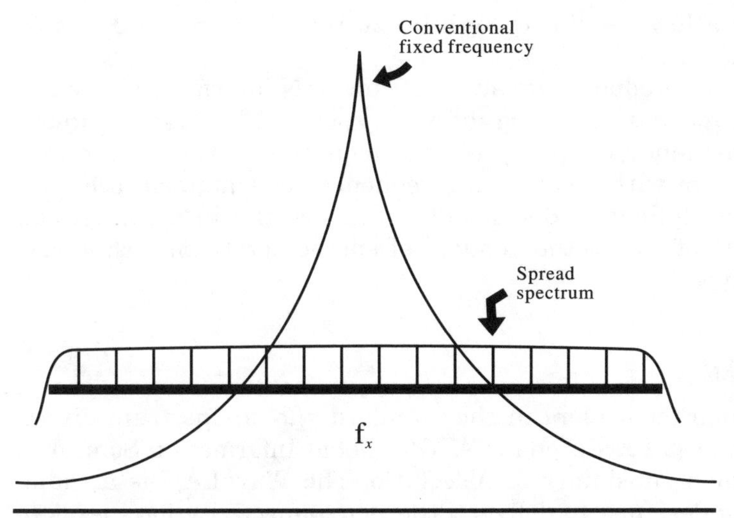

Figure 9.7 The fixed-frequency wave energy is concentrated at one frequency (f_x), whereas in spread-spectrum the energy is spread across multiple frequencies.

2. It must not interfere with other radio frequency equipment in the area.
3. It may have to go through a registration program (in the international sector, this is called homologation or type acceptance), and the frequencies may be different—using the 902–928 MHz, 2.4–2.4835 GHz, or 5.75–5.825 GHz frequencies in the various ISM bands.

Regardless of the actual frequencies being used, the concept is the same. First and foremost, use of the spread spectrum in the 902–928 MHz band makes it simpler for the vendors to build and market wireless LAN services without licensing, as is required of vendors using the fixed frequencies. Under the terms of the nonlicensed technology, the portability and flexibility of the system are easier to deal with, since a user can pick up the components and move them anywhere. Spreading across the multiple frequencies in this mode puts very little RF energy into the air and therefore minimizes the risk of interference. Since the signal is not detectable to a conventional receiver, the spread-spectrum system is less susceptible to interference from the fixed-frequency transmitters. This solution sounds workable for the transmission of an organization's critical data in a LAN environment. The LAN is therefore set up to use the spread-spectrum frequencies to allow flexibility, minimize interference, and not be license-bound.

Vendors have produced spread-spectrum LAN interfaces under two separate approaches achieving different speeds. The first is spread-spectrum frequency hopping, as described above. The second is a spread spectrum with a coded-chip sequence, as examined in earlier chapters. The following discussions compare the interconnection arrangements of two of the vendors of spread-spectrum technology, one of each type.

AT&T WaveLAN

One of the market leaders in the standard spread-spectrum direct-sequence wireless LAN arena is AT&T Global Information Solutions, a wholly-owned subsidiary of AT&T Co. The WaveLAN is a radio-based system designed to deliver Ethernet connectivity for speeds of up to 2 Mbps. AT&T uses CSMA/CA, as do all other wireless LAN services (including ALTAIR). Unfortunately, each of the wireless vendors puts a slightly different spin on collision avoidance, making the technique proprietary and not interchangeable with products and services from other manufacturers.

WaveLAN uses its own internal card, manufactured and marketed by AT&T Global Information Solutions, instead of a standard NIC card. Separate cards are used for the standard PC slots in the following systems:

1. ISA is the industry standard architecture for the 8-bit and 16-bit cards used in the 80XX slots for IBM-compatible devices.
2. EISA is the extended industry standard architecture card that supports the 32-bit slot for IBM-compatible systems.
3. MCA is the microchannel architecture card that IBM created for its PS2 line of systems.
4. PCMCIA, Personal Computer Memory Card International Association, is the interface for portable computers.

The WaveLAN card has a built-in radio transceiver, as shown in Fig. 9.8. It is the spread-spectrum capability of this card that makes the radio system work. A separate antenna is attached to the card to allow for external mounting from the PC, workstation, or other device. As mentioned above, WaveLAN supports speeds of up to 2 Mbps and will work with various network operating systems, making it essential that users understand the limitations of the support

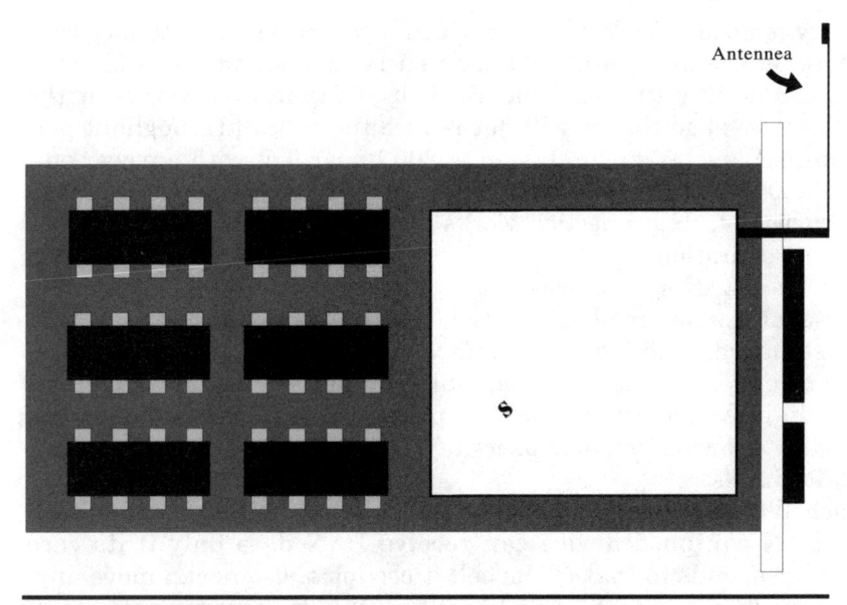

Figure 9.8 AT&T WaveLAN cards support ISA, EISA, or MCA with built-in transceivers. The antenna is outside the PC.

system in the WaveLAN infrastructure. These support systems include:

- Netware 2.X, 3.X, and 4.X series and Personal Netware (Novell)
- StarLAN/Stargroup (AT&T)
- LanServer (IBM)
- LanManager and TCP/IP via NDIS cards (Microsoft)
- Lantastic (Artisoft)
- Windows for Workgroups and Windows NT (Microsoft)
- O/S 2 (IBM)

The WaveLAN card uses a direct-sequence quadrature phase shift keying (QPSK) multiplexing technique to transmit across the entire broadband at higher signal rates. Through multiplication of the original narrow band signal, the code is spread across several frequencies. NCR also uses modulation techniques, such as QPSK, and transmission protocols that are proprietary. This approach offers better security, since the conventional radio receiver cannot decode the signal without knowing the actual spreading pattern. However, if additional security is needed, AT&T has provided a socket on the interface card to allow for the installation of a data encryption standard (DES) chip. The DES (set by the National Bureau of Standards) works at the hardware level so that it will not reduce the overall throughput performance. WaveLAN operates up to 800 linear feet with a power output of 250 mW. In part because of the higher-power output, this general-purpose LAN connection works well at most distances up to the 800-foot separation.

The WaveLAN system was not very mobile until AT&T introduced a Personal Computer Memory Card International Association (PCMCIA) standard, defining the interface requirements and construction of input-output devices for portable computers. Today WaveLAN works in any laptop, notebook, or palmtop PC that is equipped for PCMCIA, allowing nomadic users to operate in a form of cellular network for LANs.

Each WaveLAN cell is assigned its own identification code. Each WaveLAN-equipped device can receive LAN data only if its card code corresponds to that of the cell it occupies. Users can move anywhere in their assigned cell and still be able to communicate intracell. If users need to move between cells, they must first quit the application running, then reconfigure their address ID to match the cell they are moving into; with roaming this will be automatic. AT&T has released and supports an arrangement that links LANs together via a wireless bridge (called WavePOINT). As with any of the wireless LAN products today, WaveLAN is capable of interfacing directly to the backbone cable systems at standard LAN cable speeds.

Telesystems' ARLAN

Another approach to the use of the spread-spectrum technique in a wireless environment is the use of a direct sequence spread spectrum technique with a lower spreading ratio. Telesystems SLW Inc. has basically created microcell technology for the wireless LAN arena. Using a conventional cabling system, they attach the ARLAN (advanced radio LAN) devices called *accesspoints* to the cable, allowing for a full range of interconnectivity. A microcell can also be configured from the backbone network by setting an access point to act like a wireless repeater. The microcell concept is a very integral part of the ARLAN product line. First, Telesystems pioneered its microcell architecture, trademarked as Telesystems Microcellular Architecture (TMA), to allow the network to cover various applications and various-sized facilities. By using multiple base station antennas, the network can be extended to create microcells, each with its own operating area and devices.

TMA is supported by resident firmware in each of the ARLAN devices and will support multiple overlapping cells creating a seamless network within the building. The handoff from cell to cell is a part of the network concept developed into the ARLAN product line. This allows for the nomadic LAN connectivity of users who need to move freely throughout departments or floors within a building. Using the spread spectrum technology the system can select various center frequencies. This allows for the coexistence of multiple devices operating within the same area serving different needs.

The Telesystems ARLAN 100 was designed for high noise, industrial applications and has a spreading ratio of up to 100 (other ARLAN products use lower coding or spreading ratios down in the range of 11). Offering a full range of interfaces for asynchronous and synchronous data transfers from terminals and hosts, the ARLAN also has a series of wireless bridges to interconnect multiple microcells. The system operates in the 915 MHz and 2.4 GHz frequency range and uses packet burst duplex transmission capabilities.

Access to the ARLAN network is packet switched/CSMA/CA. Power output for these devices is up to 1 Watt, yielding a distance of up to 500 feet diameter in an office environment and up to 3000 feet diameter in factory or open plan offices indoors. For building-to-building communications with line of sight, the system can achieve distances of 6 miles. Using the microcell architecture each cell is capable of handling up to 1 Mbps. The range of products are shown in Table 9.5 since the offerings cover such a broad scope. In this table the product names are listed to differentiate the services or pieces.

The ARLAN 655 and 670 are complete wireless network interface cards that mount inside a PC, workstation, or other device being used. It provides the same functionality as a conventional LAN adapter card (NIC) and can support multiple topologies in conjunction

TABLE 9.5 Summary of the Telesystems Components

Product	Capability/functionality
610 Wireless Ethernet access point	Microcell controllers
655 ISA card	Interface for standard AT bus
670 Microchannel card	Interface for PS2 product lines
680 Parallel port cards	Parallel Async/Sync Data transfer
PCMCIA card	Laptop/notebook cards
620 Ethernet bridge	Bridging segments together

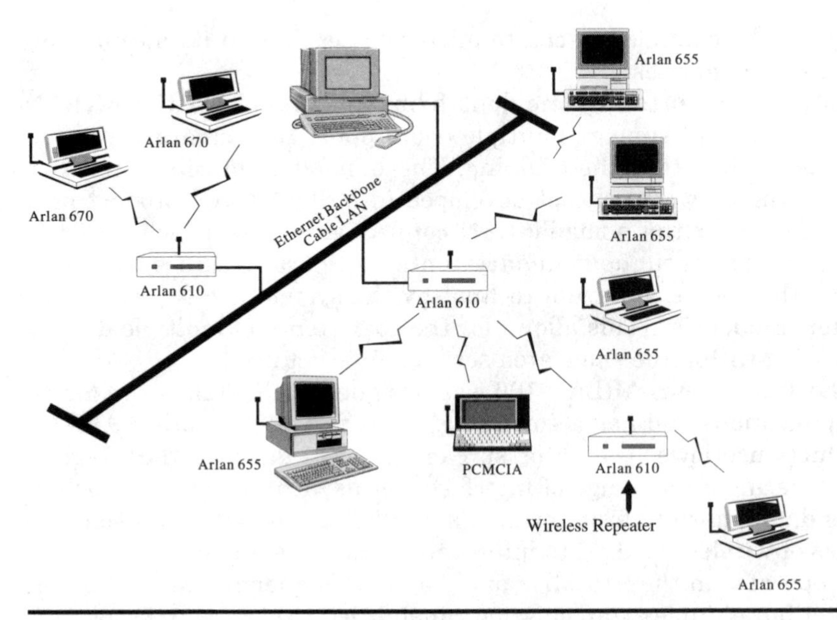

Figure 9.9 The basic building blocks of the ARLAN include the 610 hub and the ISA/PCMCIA or MCA cards. (*Courtesy of Telesystems SLW Inc.*)

with the network operating systems. Ethernet, token ring, and ARCnet are all supported with a Novell operating system. In Fig. 9.9 the basic building blocks of the ARLAN network are shown. In Fig. 9.10 two Ethernets are bridged together using the ARLAN 620 bridge.

Infrared Light Systems

As mentioned, there are two different approaches to infrared light-based LANs. The first is a point-to-point connection; the second is a diffused-light (or multipoint) service. The bandwidth and wavelength are different based on the choice made. Infrared is also simple in terms of the setup and it does not require licensing. Thus, the network is easy to acquire and use. To use the light systems the limitations will be different such as:

- Limited distances
- Line of sight for point to point
- Greater or lesser bandwidth depending on the system and choices made
- Will not penetrate walls and floors

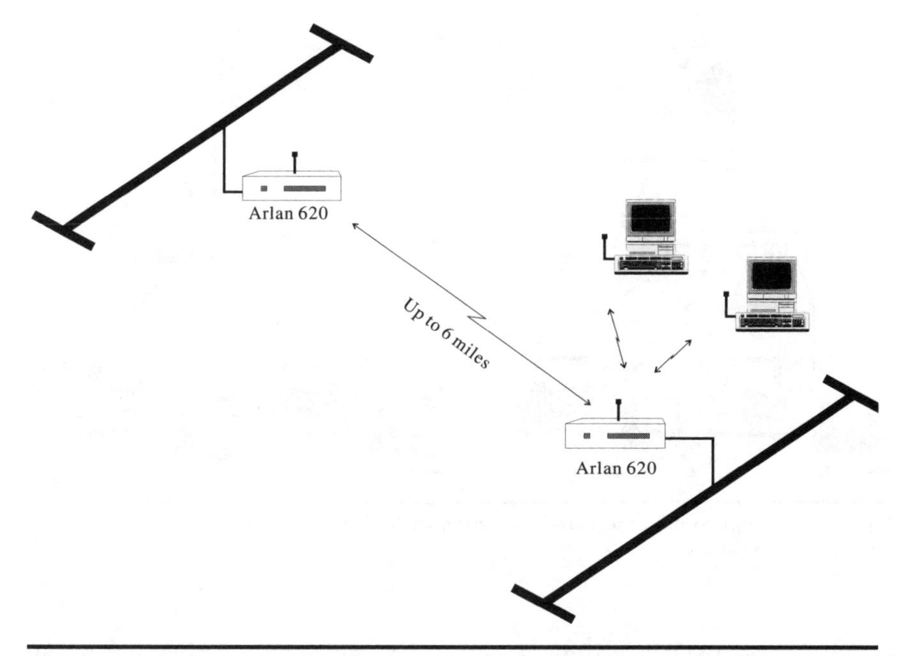

Figure 9.10 The 620 bridge can link 2 segments together up to 6 miles apart. (*Courtesy of Telesystems SLW Inc.*)

Although the limitations may sound like a detriment, the security of the data is a little easier to assure. Since the light will not penetrate walls and floors, the risk of an eavesdropper sitting outside the area and picking off the data is minimized. Further, since the light is line of sight (even with a diffused light system) any eavesdropper would clearly be visible in the area. Consequently, the use of light carries some distinct advantages. The lightwave based on 870–900 nm wavelength is not prone to the same interference problems as the radio frequency transmission systems. This again is a plus. To use the light the infrared spectrum is just below the visible light spectrum, so regular visible light will not cause a problem with the transmission of the data across the airwaves. Looking at two approaches involves the decision of cost, distances, and the speed desired. The information and vendor products listed below are for comparative analysis only.

Infralan Technologies' INFRALAN

A player in the point-to-point infrared wireless LAN is InfraLAN Technologies Inc., with the InfraLAN system. This company provides

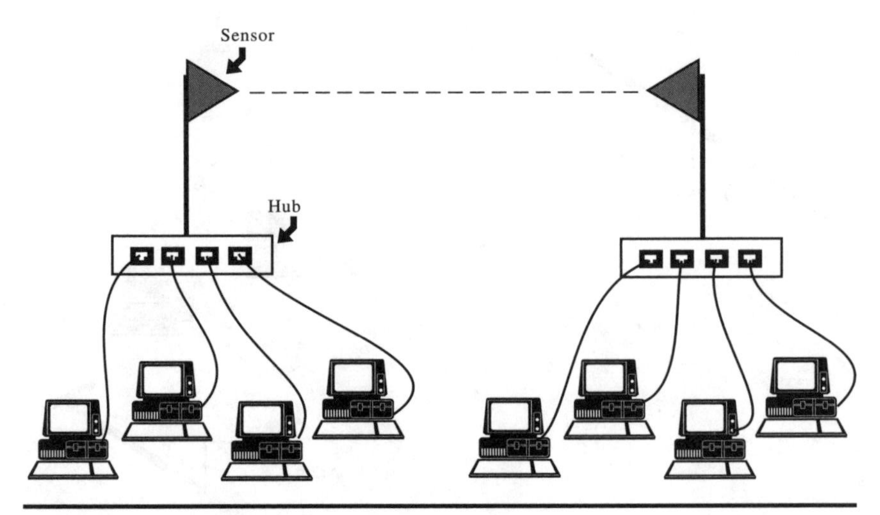

Figure 9.11 Point-to-point infrared can support high bandwidth. A standard NIC is used in the PC.

a token-ring network that is capable of supporting either 4 or 16 Mbps throughput (and on Ethernet version at 10 Mbps with true CSMA/CD) thanks to the point-to-point service. Figure 9.11 shows the infrared network. Essentially, the system will work with any 802.5* or 802.3 network operating system, and will plug directly into any standard interface card (NIC). The system operates at the physical layer (Layer 1) of the OSI reference model and is standards compliant. No special drivers or software are required for the use of this system. The point-to-point distances allowed on this system are approximately 10 to 80 feet between nodes. The system uses a base unit that resembles a standard media or multistation access unit (MAU) and two optical nodes. Two are needed to create the actual LAN. In a ring environment, a third node would be required to establish the pass through and the actual ring providing a wraparound dual path, as shown in Fig. 9.12. Each of the system components uses an optical sensor to detect the presence or absence of the transmitted light. Two sensors are required at each optical node, one for the ring-in and one for the ring-out. By sheer nature of a ring topology, the direction of one of the light paths could be interfered with and the ring would wrap around and reverse itself. This is a strong benefit when comparing the technologies. The ability to create a wraparound prevents

*802.5 is the international and IEEE standard for the token passing ring network services in the media access control layer.

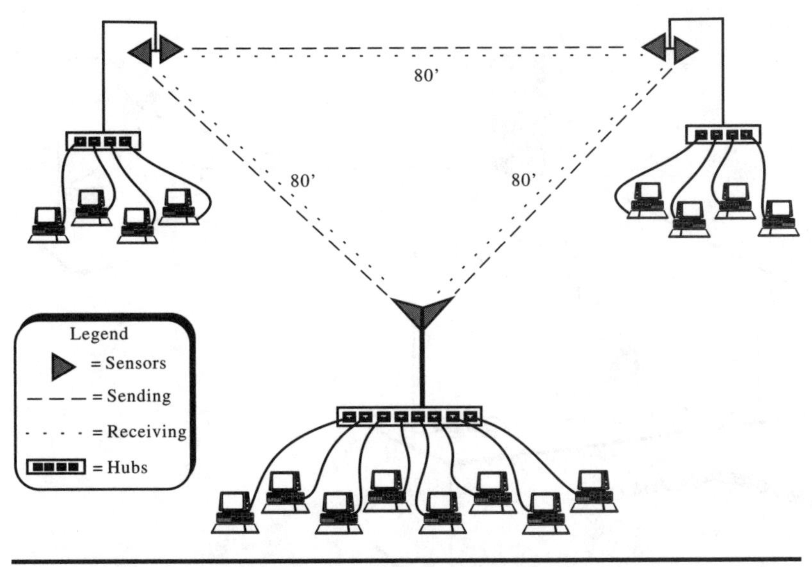

Figure 9.12 The point-to-point InfraLAN can support 4 or 16 Mbps token ring, or 10 Mbps Ethernet. Wrap-around requires 6 sensors.

downtime for a LAN that may be carrying mission-critical data. Thus the use of this concept makes sense. A summary of the actual operating systems and the drivers for the token ring cards supported by the InfraLAN are shown below. This is for comparative purposes only, since users would still have to speak with the vendor about compliance with other requirements before being sure that the system works in their environment.

Network Operating Systems	Network Drivers
Novell Netware 3.X	NDIS
Novell Netware 4.X	ODI
Banyan Vines	TCP/IP
IBM LanServer	IPX
Microsoft LanManager	

Since this network can support speeds of 4, 10, or 16 Mbps, the argument that a wireless technique cannot support the same speeds and capacities as the wired network no longer holds true. However, the configuration is set up a little differently. Using the equipment, the sensors must be mounted high in the environment to prevent an obstacle from being positioned in the direct path of the light that would disrupt the signal and the inadvertent movement of the sen-

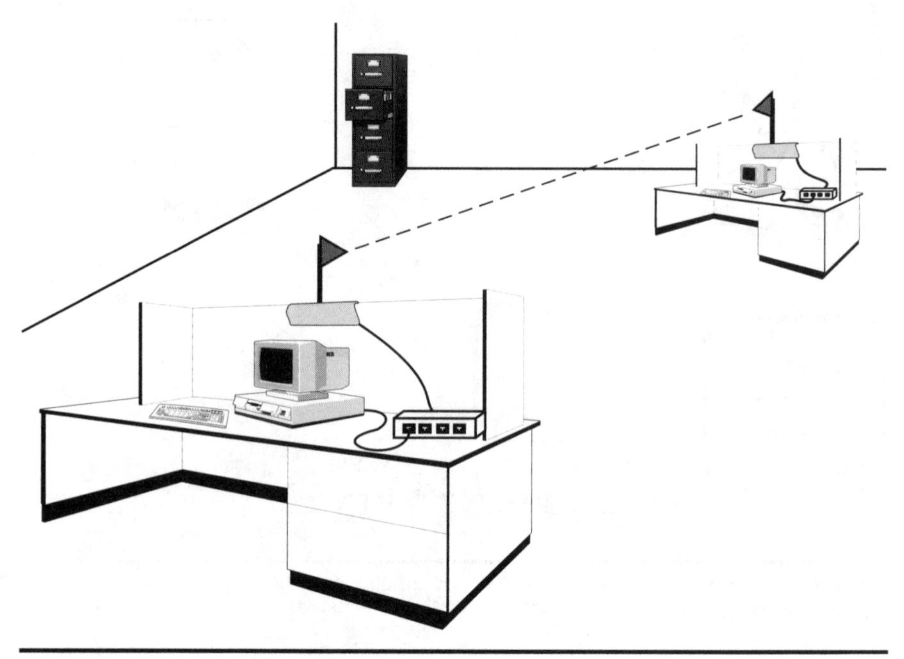

Figure 9.13 Sensors are mounted on top of an office partition.

sors that will cause loss of alignment. The sensors should be mounted to stable, nonmovable surfaces that are smooth. This will give the best results. In Fig. 9.13 the sensors are mounted on top of a partition in an office environment. The partition mount allows for quick setup between nodes at distances of up to 125 feet apart. Impressive bit error rates can be achieved at 10^{-11} up to 80 feet (10^{-9} is guaranteed by InfraLAN). The transmitter generates a conical-shaped light beam that is 5 feet in diameter at the receiving end. This facilitates quick alignment and provides reliability in the event the systems (or partitions) are bumped. The five-foot diameter of light aids in keeping alignment in both office and factory environments. The system is price competitive but not the least expensive solution on the market. However, when considering a wireless LAN, price alone should not be the deciding factor. One must also consider the throughput of the LAN, since this is one of the few wireless solutions that delivers native speeds within the environment. Clearly mobility will be limited with the use of this type of point-to-point networking arrangement. Since the line of sight is required to connect the devices, the placement of the devices will be critical. Portability of a node is possible as long as the line-of-sight can be kept. Quickly moving this system to a new location is a strong point of the InfraLAN.

Photonics, Inc.: Collaborative PC

The collaborative PC concept from Photonics, Inc. is a diffused-infrared light technology that creates instant wireless connectivity between desktop and portable PCs. The Photonics infrared transceiver creates a *collaborate,* a wireless network for indoor mobile computing. Comprised of an internal ISA (half-card) adapter card and an external-wired transceiver, the collaborative PC provides the convenience of a combined wireless connection for both desktop and laptop (notebook) computers. The wired portion of the connection comes in various forms from a clamp, an adhesive attachment, or a desktop mount to attach the transceiver to the actual PC. The system transfers data at a rate of 1 Mbps of infrared-diffused light using CSMA/CA.

Transceivers

Infrared transceivers are used for a variety of reasons: low cost, low power requirements, decent data rates, and the ability to use the light source without licensing requirements getting in the way. Many of the systems in use today are directionalized, point-to-point service provided in the form of infrared light. Photonics, Inc. chose to use a diffused light system that eliminates the need for specific point-to-point aiming. The light beams (signals) bounce off all surfaces within the office environment, such as floors, ceilings, and walls. Since this diffusion is bouncing in so many directions simultaneously, signal connectivity is more assured. The bounding effect makes it very difficult to block the signal from getting to its destination. In Fig. 9.14, an example of the bouncing effect of diffused light is shown. The light is transmitted in an omni-directional manner. The bounding of the light off various surfaces within the room therefore creates a signal that can be detected by the receiving device, such as a server, or a server that is attached to a wired LAN. The range for the transceiver is approximately 30 × 30 feet, which will not propagate beyond the walls of a contained room. If additional coverage is needed, a repeater can be installed. A single repeater will net approximately a 60 × 60 feet coverage area. This allows for some added security since the signal in light form will not penetrate through walls, ceilings, and floors. A separate LAN attachment can be created in each of many closed environments. For a PC-based system, the transceiver is an internal

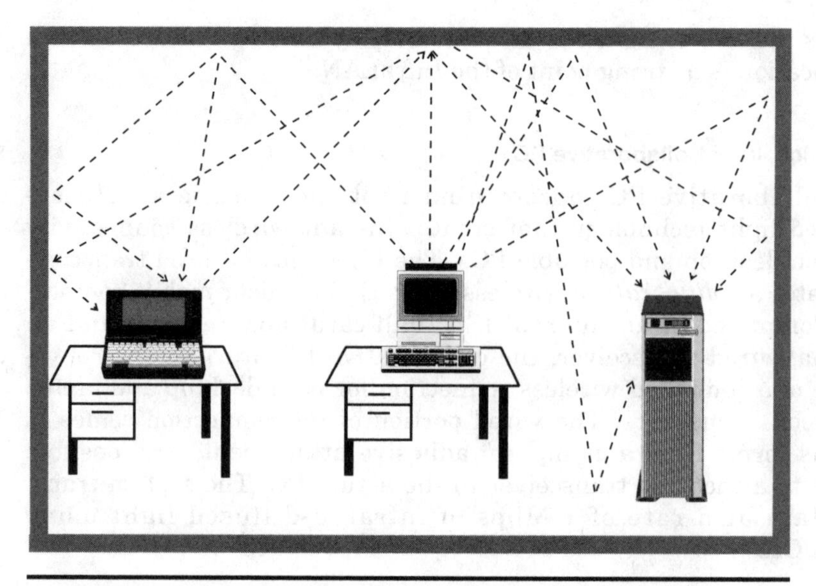

Figure 9.14 Reflective surfaces bounce the invisible light off the ceiling to the receivers.

card that fits into a standard ISA expansion slot. Figure 9.15 shows a representation of a PC with the internal card installed, and the tethered transceiver surface mounted on top of the monitor on the PC. The transceiver functions at Layers 1 (Physical) and 2 (Data Link/MAC) of the OSI Reference Model.

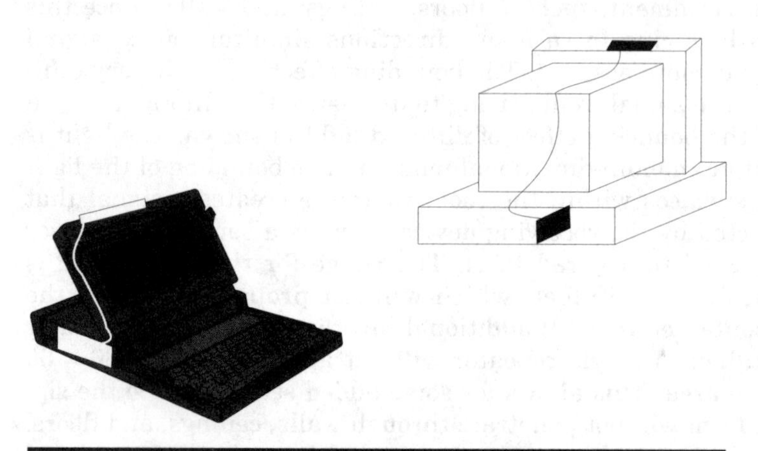

Figure 9.15 The representation of the transceiver from PHOTONICS Collaborative PC.

Network systems supported

Photonics Infrared is supported in software at the physical and media access control levels to interface with the standard Network Operating Systems (NOS) as shown in the list below. The card interfaces will represent themselves to the NOS using either an open data link (ODI) or networks device interface specification (NDIS) drivers. The future will allow an AppleTalk interface when the transceiver is attached directly to an AppleTalk port. Above and beyond the standard LAN network oprating systems, the transceivers can be used to connect to many of the operating systems that support pen-based computing, as shown below.

Photonics networks operating systems support:
Novell Netware 2.X, 3.X, 4.X
Personal Netware
Microsoft LAN Manager
AppleTalk

Other platforms users work with:
Microsoft Windows for work groups
Lotus Notes

PCMCIA attachments

Recent developments from Photonics includes a PCMCIA Type II card with a wired transceiver and a PCMCIA Type IIe, a one-piece implementation to get wireless connectivity down to the smallest of devices. The interfaces form a family of products for the standard PC, the laptop, and notebook PC, and now the personal digital assistants. This arrangement allows for true *portability* as well as *mobility* in the business arena. The infrared connectivity is fairly easy to accommodate; the lightweight design and low power requirements allow for ease of movement. The use of such devices also allows for the formation of dynamic work groups in the office environment. The dynamics of the organization that traditionally were constrained by the need to rewire the office are removed quickly through this connection.

By connecting the wireless adapters to users' devices, mobile employees can access data from a wired network through a download from the LAN server to their workstation. An added benefit is that as these mobile workers return to the office at the end of a day, they can immediately upload their data to a server or other device that is in the ready position. Ease of use, quick relocation, and price competitiveness all add to the strengths of this product.

Summary of Wireless LANs

Hopefully the user will have a better appreciation of the various options of hooking up LANs in a wireless universe based on the information covered. However, this is just a representative sampling of the network products based on each of the technology options. The differences are clear in the radio-based vs. light-based systems. The distances are fairly straightforward, and the throughputs are roughly the same for most of the wireless options. However, several vendors have emerged with spinoff products using the same concept and the same interfaces. A major issue that cannot be overlooked is the lack of standards in the wireless LAN arena. The IEEE 802.11 committee has begun formulating standards and working with the vendors to create a universal standard. However, proprietary protocols have emerged because of the lack of a standard. Thus, to develop a single standard will now take a considerable amount of effort.

Vendors who have already developed a work around that addresses their technical solution will want theirs to be the de facto standard. Yet there are at least a dozen options available to choose from, or to work around so that existing services and systems are not obsoleted. This causes added confusion and tension in the pursuit of a wireless solution to a business need. Further, the groups involved with the standards setting and recommendations are very optimistic that by June/July 1994, several of the issues will be resolved. This sounds a bit unrealistic since many of the standards will take years to develop and the vendors' vested interests must be met. With newer manufacturers entering the industry every day, the complexity of this task becomes far more formidable. Couple this with the

TABLE 9.6 Summary of Wired and Wireless Systems

Type of network	Speed	License	Topology
Wired ring	4 or 16 Mbps	No	Ring
Wired Ethernet	10 Mbps	No	Bus
Wireless Microwave	5.7 Mbps	Yes	Bus, CSMA/CA
Wireless SS, direct sequence	2.0 Mbps	No	Ring or bus CSMA/CA
Wireless SS, coded chip	.23–2.0 Mbps	No	Ring or bus CSMA/CA
Wireless point-to-point infrared	4 or 16 Mbps	No	Ring
Wireless point-to-point infrared	10 Mbps	No	Bus CSMA/CD
Wireless diffused-light infrared	1 Mbps	No	Star/bus CSMA/CA

selection process of need vs. cost, and the tendency of user to become enamored with technology as the state-of-the-art is the latest and greatest, the decisions will also have to work through radio vs. light based connectivity.

Clearly this decision would have to provide some capability to internetwork from a wired to a wireless network; a wireless to another wireless network; and a radio to a light based connection. This internetworking issue can be achieved in work around steps today. However, to create a truly wireless or nomadic LAN in the office of today and the future, the solutions should deal only with the need.

To summarize the capabilities of the wireless solutions then, Table 9.6 is shown as a comparison of the wired and the wireless technologies. This is for comparative purposes only. The variations of these options are too great to attempt to represent in a single document and go beyond the scope of this book.

Chapter

10

Other Wireless Applications—PBX, Data, Fax, and E-Mail

Chapter 9 presented an overview of wireless LAN communications of the future. Bundling other types of services together, this chapter addresses the emergence of different technical solutions that apply to:

- Wireless PBX applications and vendors
- Wireless data communications networks
- Wireless facsimile products and services
- Wireless E-mail

This bundling is applicable, since the services are all based on the user side of the LAN.

Mobility: Untethering the Workforce by the Year 2000

The concept of microcells and picocells will emerge strongly in the late 1990s, introducing products and services from the PBX manufacturers and add-on vendors whose peripheral devices will be all the telephone user needs to become as mobile as necessary. This is a logical extension of the PCS/PCN discussion in Chap. 5, the wireless connectivity across the PBX, PCS, and the cellular worlds, and the VSAT applications addressed in Chap. 8, bringing these services all together in a homogeneous network and internetworked capacity.

The use of a single telephone set to access a myriad of telephony options—whether at the PBX, in the local telephone company network, on a long-distance or international network, or at the residential location—is the foundation of a personal communications network in which a user is a single entity regardless of the device used on any network, anywhere in the world.

From there, the modem communications tools that will run similar to the telephone world will be applicable. The use of data communications devices to transport the needed data from any device anywhere in the world is a strong case for this connection. Mobility is the key to office and sales forces of the future. Untethering the worker from a single location or a single physical connection allows that work force to complete the mission of the organization with minimal back office support staff, quicker communications access to customer databases, and the capability of handling order processing on the move. This is where the industry is leading us; this is where we appear to want to go. Basic data communications is a function that has evolved over the years, with a good deal invested in training and configuring the user devices, scripting of log-on procedures, and the final access to host-based systems for file transfer, data manipulation at the host, or original data transfer in an office automation world.

From there the LAN emerged, allowing users to build and create their own data sets, databases, spreadsheets, and personal information management tools (scheduling, calendaring, etc.). The access to this information may require a data communications access to the LAN server or the individual's local hard disk in the office for either downloading information for later retrieval or uploading information for storage and later inclusion in existing databases.

Beyond the basic voice and data services, hard-copy devices such as facsimile machines still have a very valid use in the business community. Not all services can be addressed from a data terminal, since much of the information that is received for processing may well be in printed form, as in the cases listed below:

- Order forms from customers
- Original documents (purchase orders, stock certificates, signed documents, etc.)
- Pictures that have not been re-created or scanned into a digital form
- Legal documents that are required by courts or other regulatory organizations
- Maps and topographical information for agricultural business
- Disaster recovery efforts in which the information may show existing areas or information that is in paper form
- Patient records with handwritten notes from medical staff members

This list could go on ad infinitum, but the gist of the need is seen. These applications make a clear case for the ability to use portability in the facsimile world. Many of the PCs, laptop computers, notebooks, and personal digital assistants (PDAs) have a built-in facsimile modem that requires a wired telephone connection or, with modification, a cellular communications modem for facsimile. However, these devices all assume that the traffic takes on one of two different positions:

1. The data are going to be faxed into the device to be saved and viewed as a graphical document. This implies that a document may not be manipulated if it is received into a PC or notebook computer, which loses some of the versatility.
2. The document is in an electronic format already on the disk of a PC or notebook computer and will be transmitted in facsimile format to a facsimile machine. This again limits the application; therefore, a true facsimile system with its appropriate scanner is a must for the future.

The last extension of this application bundle is the messaging services to replace or to complement the existing alphanumeric pagers in use all around the world. This is not to imply that the actual need for paging services will diminish. In fact, it may well increase as the work force becomes more mobile. Therefore, instant access for two-way communications from a store and forward or real-time interactive messaging (E-mail) services will be required. Various vendor offerings have already emerged, and networks have been built. The true challenge is finding an easy-to-use, readily available network that will allow for ubiquitous communications messages in either text format or a hard-copy printout.

These are the challenges for the future application-specific networks that will have to either migrate existing services or build out newer ones. The challenge is to be one step ahead of the user demand, while at the same time have the financial stamina and backing to sustain the long lead times to support the network until the business community finally realizes or creates the demand.

Therefore, these applications must all be looked at. However, the applications will come from actually using the products and services rather than from devoting a lot of time to rehashing some of the technologies. The only technical discussion to be covered will address issues or concepts that have not already been introduced. This is where the fun should begin, as the applications are searched out and the technology is found to meet that need—as opposed to having a technology in search of an application. It is through this development stage that the needs assessment will be critical and an open mind must be kept. No single technology should be ruled out strictly because of the hype in the industry, be it positive or negative. The application is what should drive the technology. Keeping this in mind, the rest should be easy as the evolution progresses.

Wireless PBX

Just about every vendor in the private branch exchange (PBX) industry has developed or is developing plans to complement its wired system with a wireless option. Many office workers who once had fairly static day-to-day jobs now have added responsibilities or areas of coverage. Not long ago, any manager or supervisor who had to be mobile throughout the organization (or building) could rely on secretarial or clerical support to cover the telephone and take messages. Such was the normal operating procedure. Further, the mobile manager or supervisor had extended coverage within the group or department from other clerical staffers. This meant that no office person would leave a desk unless another was around to cover the phones. Even during lunch periods, the rules were laid fast that someone had to cover the telephone needs of the department. Thus, the mobile manager or supervisor was confident that there would always be someone available to answer the phone. In Table 10.1 the rules for telephony coverage pre-1980 are laid out for the benefit of those who have never been through this basic telephony 101.

TABLE 10.1 The Rules for Telephone Answering Pre-1980

Office rules for telephone coverage
1. All calls will be answered within two or three rings.
2. At least one staff member will be present at all times during business hours to answer the department's calls.
3. If a department is too small to have redundancy in the staff, all calls will be forwarded to the secretarial pool or the company's message center for coverage. A phone call will be made in advance of forwarding calls to these coverage areas. The party will advise the answering pool of the time he or she will be back to take calls.
4. All calls will be answered with a pleasant voice, in a true businesslike manner. Each answering party will use the appropriate greeting and give his or her name and the department name when dealing with customers.
5. If the called party is not readily available, the answering party will offer to take a message so that the call may be resolved as soon as the called party returns. The message will include the caller's name, company, and telephone number, the best time to call, and as detailed a message as the caller is willing to leave.
6. Upon their return, called parties will be given all message slips with the appropriate information so that they can respond as soon as possible.

The procedure may sound overbearing, but organizations were serious about the image that they presented to customers and callers, particularly if the called parties were part of management. Consequently, the rules were applied and followed up on by other management staffers, who would place test calls periodically to see if the system was working.

Somewhere in the 1980s things began to change dramatically. The personal company image became lost through improperly trained staff, poor attitudes in the work culture, or the shift in job responsibilities. Many secretarial or clerical workers inherited new duties, requiring them to be away from their assigned posts more frequently. Office automation began a new movement whereby managers or supervisors could perform more of their own duties (memos, letters, spreadsheets, etc.) through personal computers and the appropriate tools. Consequently, the staff began to move away from the personal touch.

Sometime around 1980, voice mail (voice messaging) emerged to take the place of personal contact. Since the staff was more mobile, and the trend toward downsizing began, more calls were directed to a communications tool that replaced the human. As shown in Fig. 10.1, the benefits of voice mail created only a temporary fix. In taking messages, staff people had long since stopped asking for details, collecting only the name and number of the calling party. Many of the messages were butchered to the point that they were illegible or erroneously recorded. Voice mail made it easier to take messages and guarantee

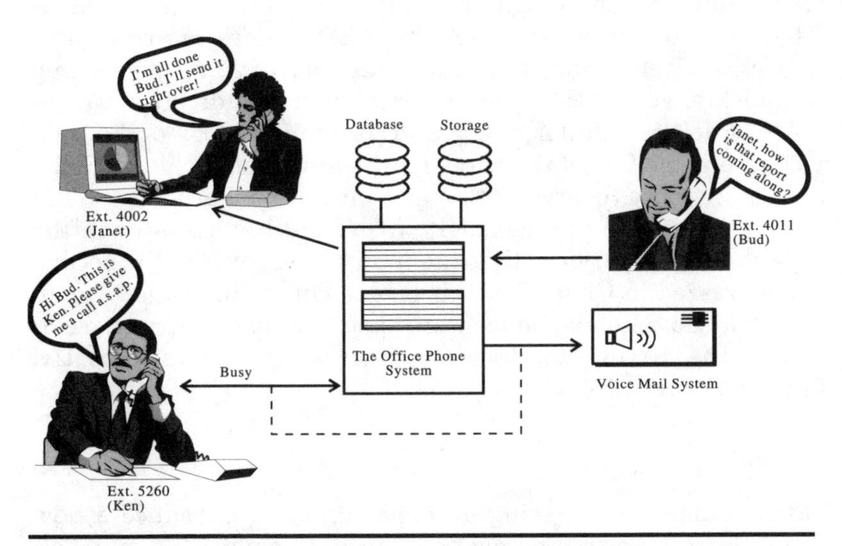

Figure 10.1 Benefits of voice mail aids the decrease of staff available to answer calls.

that all incoming calls were received. Thus, more responsibilities could be passed on the few remaining secretarial or clerical staff members. The term *administrative assistant* replaced the job function of secretary or clerk. Typically, the manager or supervisor was relegated to doing more of his or her own work.

As things continued to change and the staff became far more mobile, callers began to demand instant access to the called party. Thus, pagers were used. When someone needed to reach a supervisor or manager, the first call may well have gone to the assigned telephone number. As the process worked, the call was then forwarded to voice mail for a message. The next step was an optional one: The caller could decide not to leave a message but instead to call the pager number for the desired party. Or the caller could leave the message but mark it as urgent. Upon receipt of an urgent message, the called party could direct the voice mail system to automatically outdial to the pager. Regardless of the choice or option, the called party would still have to find a telephone and place a call back to the voice-mail system or directly back to the caller. Either way, the process was far more cumbersome than it had ever been.

To solve this problem, the industry responded with the replacement to the pager, or a supplement to it. The industry responded with the next evolution of portability, that being the cordless phone. The cordless had its pros and cons. The most significant cons included the limited distances covered from the base station, the interference problem, and the risk of being monitored. Since this was public domain air space, there were not many options available to overcome the problems. Therefore, users began to lean on the vendors to arrive at a solution. Cellular telephones were introduced as a means of reaching a person quickly, at any time no matter what that individual was in the process of doing. Cellular, being a licensed technology and a public service, overcame two of the three most significant problems of the cordless set. Actually, everyone thought that a mobile phone would overcome the monitoring problem. (It wasn't until much later in the development of these services that the industry uncovered the security issues addressed in Chap. 7 of this book.) This sounded good to all concerned, but the cellular phones were expensive both to purchase or to rent, and the airtime for incoming messages was quite costly. Therefore, something else had to be done.

Ericsson DCT 900

The PBX manufacturers, seeing an opportunity to introduce a new service and hopefully spark new sales in a sluggish PBX market, began the research and development phase of the wireless telephone that would be a part of their system. To overcome the lag in the development stages, many of the PBX manufacturers decided to use an add-on peripheral system coupled tightly to the PBX. New markets developed as either proprietary products or add-on systems were introduced quickly. However, it was not until the late 1980s and early 1990s that the reality of a wireless set for the telephone systems was evident. One of the first to introduce anything in the PBX market was a manufacturer of various communications products, L. M. Ericsson with its DCT 900 wireless telephone set for integration with its MD-110 PBX product line. It was through the pioneering efforts of Ericsson in the international communities that Ericsson made much of the headway in developing its wireless telephone module and set.

Some of the experiences that this organization had in the planning and implementation of cellular equipment around the world led to the quick development of new interfaces. Ericsson's approach is the DCT 900 portable cordless telephone using the CT-3 standard with a total of 4 MHz of bandwidth. The 4 MHz bandwidth will be broken down into 1 MHz channels that are then allocated in TDMA/TDD, as already described in Chap. 5. What the system is designed to do is use a series of picocells* spaced throughout the building in an overlapping pattern. The DCT 900 with cordless capabilities will allow for the TDMA/TDD access methods. The system uses a dynamic channel allocation (DCA) technique to avoid the necessary frequency-planning and cell-planning studies required with older versions of cell design in an FDMA/TDD methodology. Therefore, all cells can use the same frequencies and channels so long as they do not interfere with one another. To make this happen properly, the system will allow for the handset to monitor all channels simultaneously and select the one that is the strongest. While a base station is in the idle mode, it is constantly communicating on a spare channel to let the handsets know of its existence.

The DCT 900 is designed to work with the PBX to allow for the direct incoming and outgoing calls within the business, and eliminates the need for voice mail, paging via external paging systems, and the call-back procedure. Power for the handset is provided by a battery system that allows approximately 16 hours of standby time or 4 hours of talk time (or a range of combinations). The handset is small and lightweight. The dimensions are 53 mm × 155 mm × 25 mm and the weight is less than 200 cm (just under 7 ounces).† The

*Picocells are cells that will range up to 150 feet for the smooth and orderly handoff between cells located within an office building or campus on the MD110 PABX or in later versions the 150 for smaller organizations.

†Note the references to international measurements and weights. Since Ericsson is an internationally based organization, it uses the international weights and measures for its descriptions.

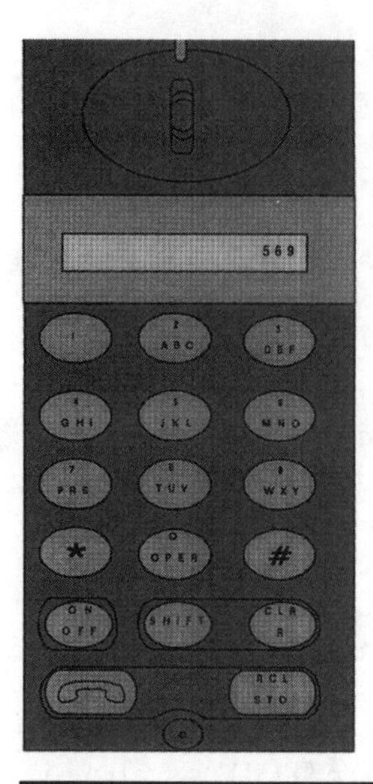

Figure 10.2 The DCT900 set from Ericsson.

original European system was designed to work in the 800 MHz frequency band, and also works in the 900 MHz range as well for U.S. implementations. The system has been enhanced to operate as the European DECT standard in the 1.9 GHz bands. Figure 10.2 is a representation of the DCT 900 telephone set. The set has 17 buttons to include the numbers from 1 to 0, the # (United States refers to this as the pound key) and * (United States refers to this as the star key) along with others such as store, recall, clear, and send keys. The set includes a 12-digit display and a paging feature that can store up to 20 messages for later retrieval. In Fig. 10.3 is a layout of the way the system would be set up. This is not much different from any design already being considered by other manufacturers. Coverages will be accommodated by unshielded twisted pair cable runs to the picocells dispersed throughout the building.

Spectra Link 2000 Pocket Communication System

One of the first players to emerge as a provider of an add-on module, though not the first to develop the technology, is a company called Spectra Link, Corp. with its Spectra Link 2000 set. Part of the hold-

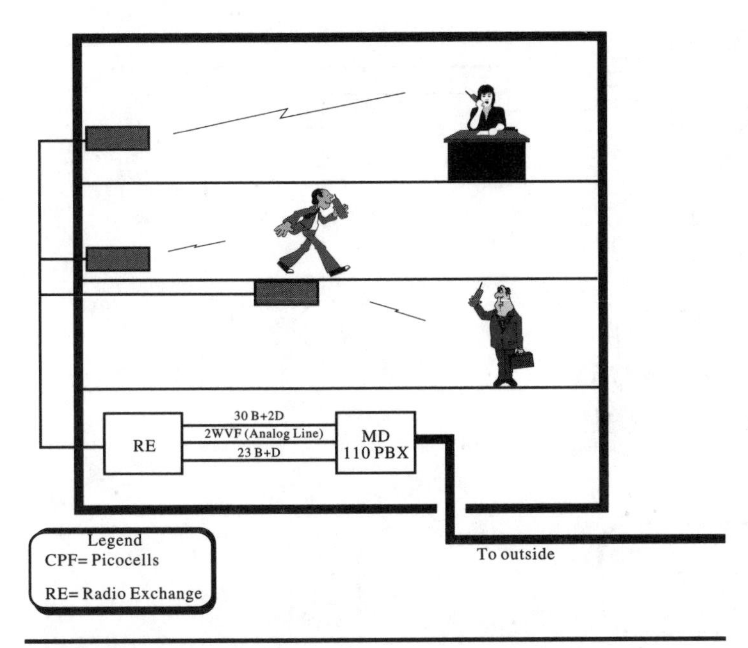

Figure 10.3 A block diagram of the MD110 & DCT900 cordless systems linked together.

up on the introduction of this service was in determining what frequency band to use: Should the vendors request licensed frequencies in an already congested and sparse spectrum, or should they opt to use the ISM unlicensed bands in the 902–928 MHz, 2.4 or 5.7 GHz frequency bands? The issues meant that, although unlicensed, the ISM bands still have type acceptance and permission requirements that must be met. To that end, most of the vendors chose to go through the unlicensed route, using wherever possible the 902–928 MHz frequency bands. This introduced some delays, but probably nothing in comparison with the delays that would have been experienced in finding frequencies to license. Spectra Link was formed in January 1990 and the initial system announcement was made in June 1990.

To accomplish the integration or coupling to other systems, the Spectra Link PCS 2000 was designed as a standalone box that would use 2 wire-loop start lines between the PBX of any manufacturer and the Spectra Link system. All that was required was an inband signaling capability as the telephone system (PBX) passed a call from one system to the other. This could be accomplished rather easily. Figure

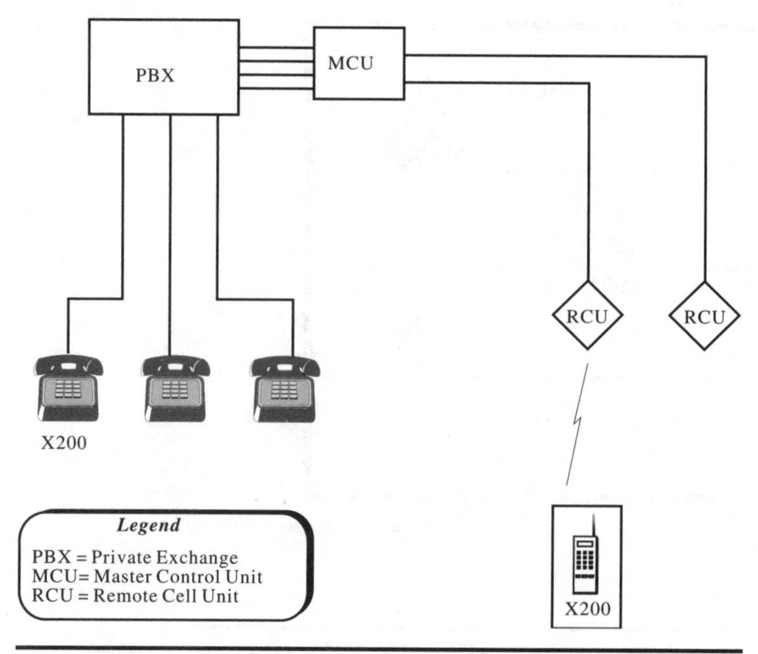

Figure 10.4 The Spectra Link PCS 2000 adjunct system connects to most PBX systems.

10.4 is a graphic representation of the connection of the PBX and the add-on module interfaced together. The Spectra Link PCS 2000 will support a large base of users in an office building or campus environment using a bandwidth of 26 MHz in the 902–928 MHz band. The base system characteristics are summarized in Table 10.2, which looks at the system in terms of numbers of users and preliminary coverages. This table information is something that will continually change as the provider enhances its product line and creates more compact capabilities.

TABLE 10.2 The Spectra Link PCS 2000 Capacities

720 telephone sets, known as the pocket telephone 2000
120 simultaneous calls in a fully configured system
180 remote cell units at maximum configuration
Integrates to most PBX manufacturers' systems
Uses CDMA/TDMA spread spectrum
Up to 3,000,000 square feet of area
Set weighs less than 6 ounces

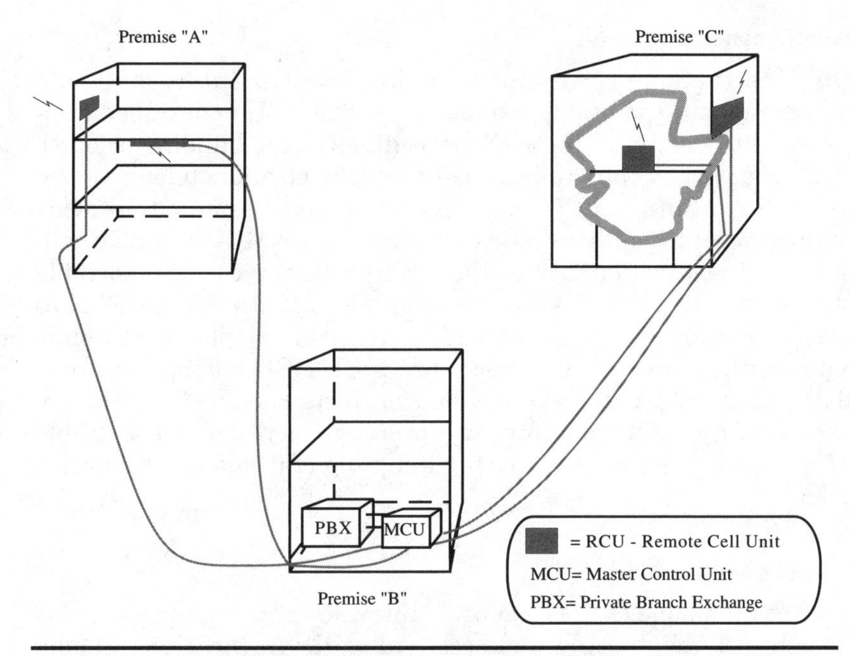

Figure 10.5 The Spectra Link system can connect multiple buildings together serving up to 720 users.

In Fig. 10.5 a block diagram of how the system will work is shown. In this drawing the system is installed adjacent to the PBX and tied together with a 2 wire loop start connection service. The basic components of the Spectra Link PCS 2000 are as follows:

The master control unit

The MCU is the brain of the system that is attached to the PBX or Centrex* system. This portion of the system is responsible for the establishing of calls to the portable sets. The MCU is hard-wired to one or more remote cell units and coordinates the servicing of dial tone (equivalent) continuously to the active portable sets. Since the MCU is designed around integration to the existing system, obsolescence is not a concern. The systems work together to provide the necessary interconnection and connectivity.

*Centrex is a partitioned portion of a central office at the local exchange. This partition acts as the PBX would. The differences in Centrex service are that the user does not have to buy PBX hardware to provide telephone service. It is a rental or lease arrangement with the local telephone company.

The remote cell unit

The RCU is the small cell site that provides the interconnection spread throughout the building or campus. The RCU is designed compact enough to mount on the walls or ceilings of the building or office space. This device is the interface between the portable telephone set and the MCU. Multiple RCUs are supported on the system to provide the seamless cell to cell handoff. Working with the MCU the RCU will provide the full functionality of the PBX or Centrex to the portable telephone set. This includes the features associated with the PBX or Centrex in a transparent manner. If the existing telephone system is equipped with a voice-mail service, the MCU/RCU combination can hand the call forwarding into voice mail as the PBX would do. Additionally, the RCU monitors the presence of the active portable sets to determine the location for the handling of incoming and outgoing calls.

The pocket telephone

The pocket telephone (PT) is the basic interface at the user level. The lightweight set weighs only 6 ounces and works in the system using full digital transmission spread-spectrum capabilities. The set has a standard 12-button keypad, plus an extra button to access the standard features of the PBX or Centrex. Internal and external calls are handled the same way on the portable set as on the standard wired desk set. Figure 10.6 is a representation of the portable telephone set.

For visiting users or traveling users the Spectra Link system will accommodate the logon for the sets as users move between offices or buildings across the country.

Other PBX vendors

Clearly every PBX manufacturer is either experimenting with or has already announced its intent to deliver the wireless adjunct or integrated capability. These are all based on some of the development stages that have been committed to. To quickly summarize the efforts of the PBX vendors, they would include:

Northern Telecomm Inc. With its Companion 10 and 100* systems, Northern Telecomm is currently in trial on an experimental agreement from the FCC because it is using the 864–868 MHz frequency

---

*The Companion 10 and 100 are registered trademarks of Northern Telecomm Inc., the model 10 will support up to 32 portables and the model 100 will support up to 80 portables. Future developments will be expected.

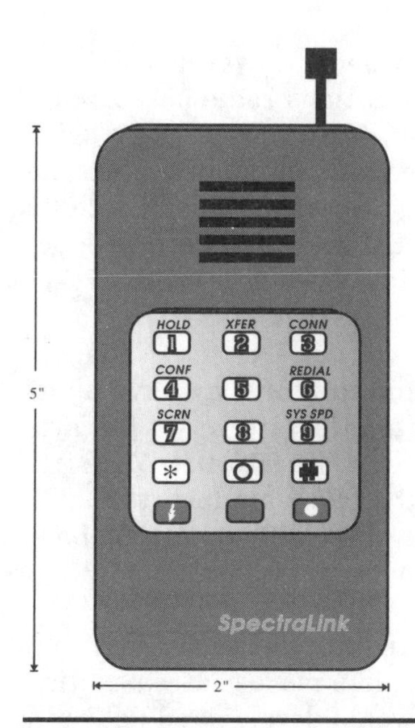

Figure 10.6 The Spectra Link PCS 2000 portable set (actual size, 5″ × 2″).

bands which were authorized for a trial of PCS by the FCC. Under the terms of the experimental arrangement NTI was allowed to test only, not market, up to 100 systems. NTI got their hands slapped because the FCC got word that NTI had exceeded the installation of 100 users, and refused to deliver a list of the pilot locations. This appears to have been resolved and the permission is forthcoming. In 1994 NTI will likely be offering a commercial product.

Intecom, Inc. Intecom has announced its support and delivery of the adjunct using the Spectra Link PCS 2000 rather than undertake their own development. The Spectra Link will integrate with the IBX and Telari* products in a seamless fashion. Commercial installations have already been delivered and tested. Intecom's primary market penetration is in the medical and educational field. Their trials have been fairly successful from all indications.

NEC. This company has been delivering adjunct boxes to the 1400 and 2400 series PBX systems in the past. The adjuncts used the

*Integrated Branch Exchange and Telari are registered trademarks of Intecom, Inc.

Spectra Link PCS 2000 in the initial deployment. However, NEC is a manufacturer of a wide variety of communications radio-based products. Therefore, it will likely be bringing out its own proprietary product (or adjunct) in 1994. This is not commercially announced, but one can expect to see such a development. The Spectra Link PCS 2000 series worked and was available before NEC had their own product ready. Now that NEC has had more time it is only natural to expect them to deliver their own product.

Rolm/Siemens. Rolm/Siemens has announced that they have plans to deliver in early 1994 a wireless product to integrate with their system. This system as an adjunct will operate in the 902–928 MHz frequency band. The system will work with the 975X product, the 9000 and 8000* CBX lines made by Rolm. The system will support 80 base stations, each base station supporting up to 8 conversations.

AT&T. AT&T was late in announcing their introductions and intentions to develop a wireless adjunct. However, in a move that was anticipated, they finally did announce that they would develop and deliver a product for their Definity and Systems 85/75 series. However, they chose to look at developing in the microwave frequencies that they already have licenses to (6 GHz band). They also have a key system arrangement for their Merlin and Partner† plus line.

Clearly one can see that different approaches have been taken by the PBX manufacturers, and that each has used a slightly different approach to the frequency bands supported. The proprietary nature of their offerings, with the exception of the adjunct boxes, will require that careful attention be paid to the integration of these technologies in an office environment. Further, now that the FCC has allocated bandwidth in the PCS spectrum, many of the vendors may opt to move into that band, allowing the portable PBX handset to be used in the PCS and the PBX worlds simultaneously. This may mean that the networks will have to support distinctive ringing arrangements to separate or differentiate the incoming calls from the office vs. the personal communications or residential calls. More development will follow in this regard over the next 5 to 6 years so that a single integrated device will be usable on any network as either a personal or a business phone. The lines of demarcation and distinction will then be drawn for the ubiquitous communications capability anywhere to anywhere.

---

*The 9751, 9000 CBX, and 8000 CBX are registered trademarks of Rolm Corp.

†Definity, System 75 & 85, Merlin and Partner Plus are all registered trademarks of AT&T.

Wireless Data Communications

This is a subject that is possibly the hottest in the industry today. The use of a wireless data communications technique has literally opened the doors from the confines of the office and made the work force far more mobile and versatile at the same time. The development of wireless data transmission began with the introduction of the cellular industry. The very simplest of data needs could not be met because of potential problems in the transmission path which will be discussed later. Mobile communications suppliers had to find solutions to the transport of data communications if they expected support and sustained use of their networks.

Otherwise if a user had to leave the vehicle to place a data call on the wired communications network, then a voice call might just as well be made at the same time. As the sales and marketing staffers were put on the road, and cellular telephony was provided, only a small percentage of their needs were met. Still necessary was the requirement to access inventories, databases, and other information such as calendars that were back in the office. In order to access these services, the mobile user attempted to gain access via a dial-up communications service. However, some distinct differences exist in modem communications. First, a modem is trained to wait for dial tone. However, in a cellular world, there is no dial tone per se. The user dials the number and presses the Send button, which uses RF to transmit the information. Therefore, the modem was rendered somewhat useless, since it did not see the dial tone or receive any DTMF* progress tones across the network. Therefore, in order to even get anything to function close, a work around had to be devised. This meant that the user could manually dial into a data line modem at the receiving end. Upon hearing the tone from the receiver, the user would then start a transmit tone to create a handshake. This was cumbersome at best.

Another problem began to plague the mobile worker. As data were being transmitted across the cellular network, the possibility of loss of carrier existed when the handoff from cell to cell took place. This meant that the connection could be lost in the middle of the transmission, requiring the user to start all over. Frustration led the way for those strong-hearted individuals who attempted to use the cellular voice network to transmit data. To overcome some of these problems the carriers and vendors of products developed a group of cellular modems. These were helpful but the limitations were still evident.

*Dual tone multi-frequency tones or touch tone service is the service where the progress tones can be sent inband. Touch-Tone is a registered trademark of AT&T.

TABLE 10.3 Summary of Data Communications Characteristics on a Cellular Network

Data communications on the cellular network
1. Was limited to 2400 Bps
2. Could not be handled through a handoff from cell to cell, loss of carrier would ensue
3. Expensive to use and to keep current
4. Transmission impairments common on RF networks could severely limit throughput (i.e., latency, noise, static, fade)
5. May not be compatible with standard wired data modems, due to proprietary error correction and compression techniques
6. Lacked robustness and ubiquity

Specifically, if asked today, many cellular users would indicate that data cannot be transmitted over a cellular network. Fortunately, this is not the case any more. In Table. 10.3 a summary of the data communications services are provided. These include the initial services offerings from the carriers.

Some of the applications for the transport of data, so that an understanding is achieved as to why bother, are highlighted below in Table 10.4. This is a partial list of the forms and potential users of the wireless transmission of data. The need for instantaneous data may be the need, the ability to access the data from a far remote area can be an application. However, these can be broken down into the categories in this table.

These applications are in use by many today. However, there are still problems with the actual transmission going through. The author is not a proponent of persons driving down the road and attempting to transmit information at the same time. This is represented in the drawing of the connections available in Fig. 10.7. However, as a vehicle may be used to do just that, the handoff from cell to cell can become a problem. The reason is the functionality in a modem. The limitation of the 2400 Bps and the not handoff were functions of how a modem looks at continuous carrier tone between the send and receive modems. A quick tangent here will help to clarify this point for those not familiar with how a modem works.

- First when the modem is set to communicate, it uses transmission protocols that check the line for the dial tone. When dial tone is detected, the modem will go through a dial sequence to the called number.
- Next, the receiving modem detects ringing current on the line, goes off hook, and provides a high-pitched tone to the line. This high-pitched tone is designed as a recognition tone. Each type of modem

TABLE 10.4 Some of the Applications for Wireless Data Communications

Applications for wireless data communications
Fire departments that need information on chemicals that can be used in fighting a blaze
Police departments running a file on a suspected felon
Fire departments accessing information from a location regarding the construction of a building, underground tunnels, or access and egress systems, etc
Emergency response teams obtaining information on call out numbers and lists of reaction teams. Further information on sequencing of events that may be contained in a disaster recovery team
Retail organizations running "sales" in their parking lots, and communicating credit information or inventory information on customers
Field service personnel accessing parts and inventory information from a database, and placing orders for the parts to be shipped from inventory immediately
Emergency rescue personnel transmitting information such as EKG or EEG output to medical facilities
Trucking organizations that may send bills of lading, or rerouting information to the vehicle to save time and money. The use of new vehicle route directions to dispatch a vehicle around a problem area.
Banking organizations that cannot afford to install wires to the localized or remote links to ATM* machines.
Sales forces accessing information on current pricing information prior to committing a sale or a delivery date.
Medical staff providing patient tracking information, or using a messaging system to schedule operating room availability.
Delivery organizations tracking parcels along the delivery cycle, to provide instant information to the recipient or the sender.

*ATM refers in this situation to the automated teller machines.

emits a different high-pitched tone. That is, a low-speed, high-speed or a fax modem each have a different tone sequence. Upon recognition between both modems that they have the same capability to transmit/receive from each other, they then go into a handshake mode. If by chance the tones between the two devices are wrong, then the call will time out and the line will be dropped.

- Once the handshake begins, the modems swap transmission protocol information on the speed they will be transmitting at, the number of data bits, the number of start and stop bits, the use of parity bits, etc. Assuming they come to agreement, a carrier detect (CD) light goes on. This carrier detect is a constant unmodulated signal that is being sent from the modems at all times. If the carrier disappears for a certain period (80 ms) than the modem can assume that something has happened to the other end and will terminate the call by dropping the line.

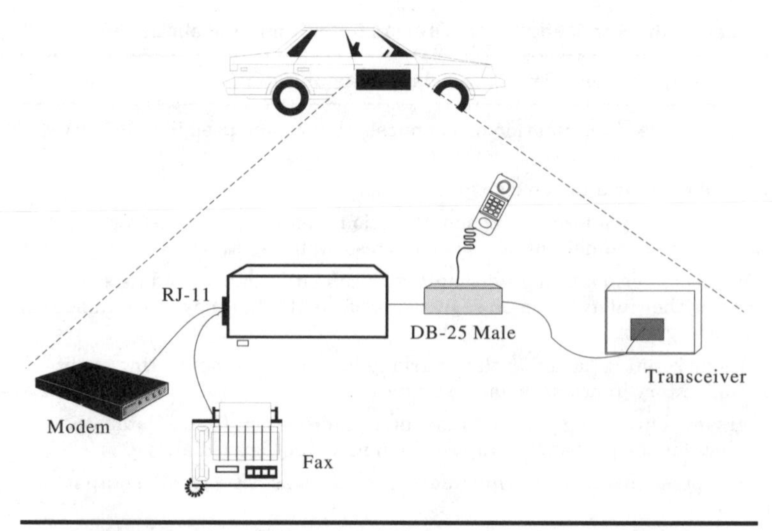

Figure 10.7 The connec tions possible in a vehicle for wireless data communications.

- As the data transmission is taking place on a cellular network, the problem arises as handoff occurs. There are other reasons for termination of calls but this example is used to show some of the problems. As the handoff from one cell to another occurs, the loss of a carrier will take longer than the system will allow. Therefore the call will be dropped due to loss of carrier. This is shown in Fig. 10.8 where the carrier tone is lost for greater than the modem will allow.

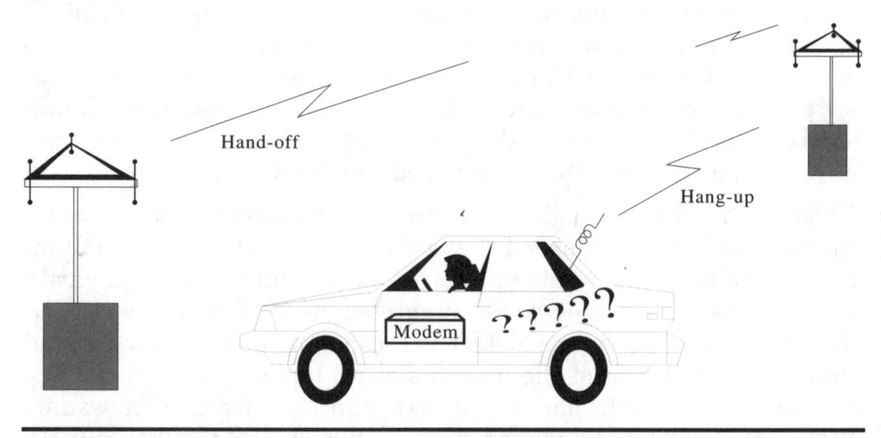

Figure 10.8 Early implementations of wireless data caused disconnects during cell-to-cell hand-off.

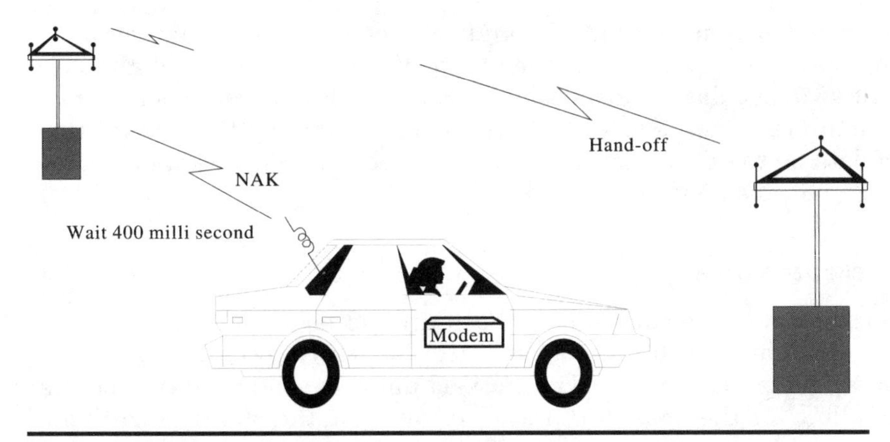

Figure 10.9 Improvements in the data communications prevent the disconnects as handoff occurs.

- To overcome this problem, the cellular modem manufacturers created a more tolerant modem that will wait for the carrier to return for up to 400 ms. This is similar to a fax modem that is far more tolerant to line hits. Thus, the modem may get a small amount of transmitted data, and produce a "NAK," but the connection will remain. This is shown in Fig. 10.9.

Motorola, one of the initial leaders in cellular communications, came up with a cellular modem that would sustain up to 9.6 Kbps of data transmission on a cellular connection. A recent development by Motorola UDS is a 14.4 Kbps data/fax modem for cellular transmission. The PCMCIA card (modem) will work with a standard laptop or notebook computer and a Motorola Micro-TAC cellular phone. The system is fully compatible with CCITT Group III fax services and can support either a wired data or a wireless interface for transmitting. Many other manufacturers have either created their own or licensed the technology.

Telebit Corp. on the west coast has developed a trailblazer (called cell blazer) modem that can transmit cellular data at up to 19.2 Kbps. Obviously what has been occurring is an evolution of the technologies and the products. Remember that the cellular channel in the network is frequency modulation with 30 KHz of bandwidth. Therefore the transport of these speeds would not be unreasonable.

The above scenarios deal with a circuit switched data communications service. This is both expensive and wasteful when evaluating this service. The cost per minute on a cellular network is between \$.45–\$.60/minute rounded to the higher minute if a fraction is used. The transmission of a typical data stream may be only thirty to forty seconds long, but the initial minute charge is applied. This works out to approximately twice to three times the cost of a wired dial-up circuit switched data transmission. Consequently, the carriers and manufacturers began to look for better ways of handling the transmission of data on the networks. To do this the use of packetized data was the next logical step in the evolution.

Packetized Data

Packet data transmission has been around for a long time. Packets are designed to utilize the circuits better than circuit switched data. A standard form of packetized data transmission is essential to allow for the ease of connectivity and for the standardization of the packets that are transmitted. In this vein, several vendors formed a consortium to establish what is known as the cellular digital packet data (CDPD) standard. The CDPD is conceptually a standard that will allow for an open-air standard so data transfer could be effective and to lower the cost of the mobile equipment and transmission expenses. The consortium is made up of the following major players:

- Ameritech Cellular
- Bell Atlantic
- Contel Cellular
- GTE MobilNet
- Mc Caw Cellular
- Nynex Mobile Communications
- PacTel Cellular
- Southwestern Bell Mobile Systems

A summary of the efforts of the consortium in the development of the CDPD standard is shown in Table 10.5. Recent events have thrown a curve at the CDPD standard with Motorola claiming patent infringement on several of their existing standards in the cellular network. Motorola is willing to license product manufacture to anyone on a non-discriminatory basis, but does require the license. This is apparently an oversight by the consortium on the development of the standard as they overlooked or chose not to evaluate Motorola's claims to the patents before the standards work was nearing completion. Consequently, vendors who were developing products to work in the CDPD environment have experienced delays or have announced that this could raise the price of the products they will be manufacturing.

TABLE 10.5 A Summary of the CDPD Standard Based on the Consortium Efforts

Cellular digital packet data standard
1. Developed by IBM as an open specification that defines how to transmit packetized data over the existing circuit switched cellular networks.
2. Enables data to be transmitted on dedicated 30 KHz cellular channels or inserted and interleaved on idle voice channels.
3. Establishes a data packet that is 128 bytes length per packet, at speeds of up to 19.2 Kbps.
4. Supports both the implementation of OSI and TCP/IP addressing schemes.
5. Allows for a usage-based pricing scheme rather than a fixed cost per minute. The pricing will be on the per packet base.
6. Supports the necessary flow control, sequencing schemes, error control and recovery and error correction techniques.
7. The user equipment can include both a telephone and a CDPD function in the set through a standard interface.

One more kink in the overall wireless armor for an easy-to-obtain standard system connection.

The use of a cellular network to transmit the equivalent packets of data should improve the total overall performance of the network as well as the integrity of the data. This is a modified version of the X.25 international standard for packet data transmission. Other suppliers have experimented with the transmission of interleaved packets of data on the cellular network. One such experiment included the ability to use a narrower band for voice communications and add the interleaved data packets on the idle channel capacities. An example of this is shown in Fig. 10.10 where a packet modem manufactured by Mobile Datacomm uses the data over and data under voice communi-

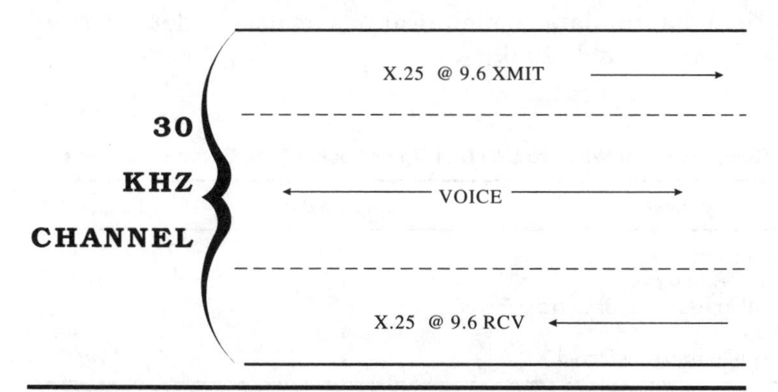

Figure 10.10 Cell data's X.25 capability over and under voice.

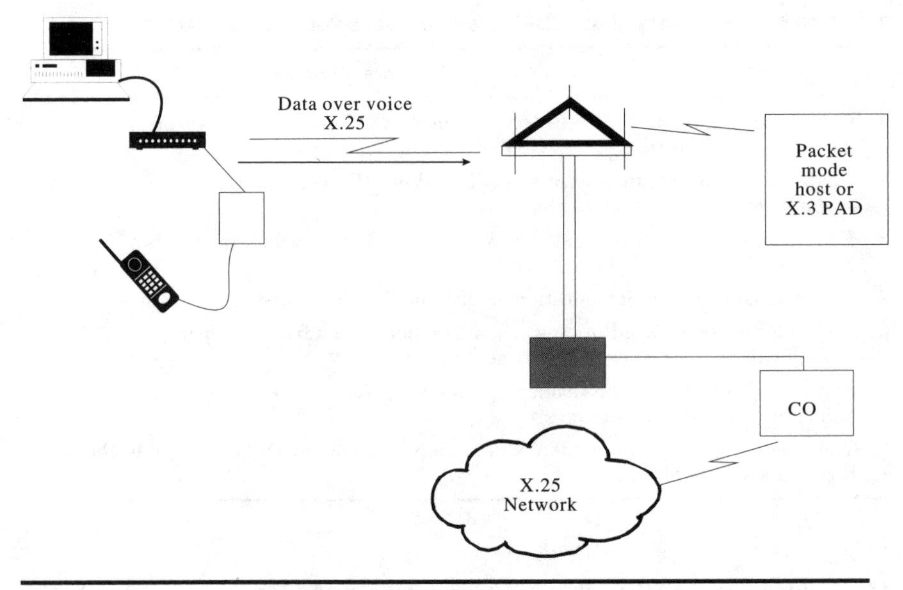

Figure 10.11 The split off of data or voice over a channel for dial-up or packet transfer.

cations by splitting the bandwidth into the equivalent of three narrow band paths. This is on a point to point basis, it was not designed to have separate paths between different locations. This arrangement is shown as three separate channels when in fact it is a single channel but interleaved transmit packets over the voice at 9.6 Kbps of X.25 packets and data under voice interleaved at 9.6 Kbps of received X.25 packet data. However, if the capability exists with the carriers, this can be split into different paths at either the cell site or the MTSO. This is shown in Fig. 10.11. A summary of the differences between using real time dial-up data communications or packet data communications is shown in Table 10.6.

TABLE 10.6 Comparison of When to Use Dial-Up or Packet Data Communications

Application	Dial-up data	Packet data
Real-time connection	Yes	Yes
Interactive data transfer	Yes	Yes
Relatively small amounts of data transferred	No	Yes
Long call set-up	Yes	Yes
Large amounts of data transferred	Yes	No

Facsimile Transmission

The use of a facsimile (fax) machine or a card in a PC device has been around in the wired world for years. This is not new technology, merely a new application for the technology that already existed. It was only a matter of time until users in vehicles would eventually want to use a fax in the vehicle. Supplementing the voice and data transmissions with the capability of using a facsimile machine in a vehicle was a natural development. Facsimile has been used effectively by the trucking industry for some time now. In this particular application the trucking organizations can ship information directly into the cab of a vehicle without too much ado. Bear in mind that truckers (drivers) are just that—they drive for a living. The use of high technology devices such as data terminals doesn't come all that naturally. Therefore, the capability to deliver a printed form into the vehicle makes it more palatable for the driver to use the information. Specifically, when using a vehicle-scheduling program, requirements to redirect a truck to a new route or location can be done via a text-based data transmission. But to make it simple, the use of a facsimile machine allows a map with a new route plotted to be delivered to the driver. Further, if a reply or change is required, the trucker can use a pen or pencil to mark up a document then ship it back via facsimile, rather than trying to type or fill in a form on a computer screen.

This application goes one step further. During the Gulf Crisis trucks crossing an international border, particularly into Mexico and Canada, were required to have new paperwork in order to cross the border. Much of this additional paperwork was the result of the border patrols closing down on the free passage of vehicles for fear of terrorist bombings, etc. Consequently, during this crisis, the trucks may have already been en route before the new forms were instituted. Arriving at the border crossing, the truckers were not allowed to proceed unless and until they got the appropriate forms filled in and delivered to the border patrols. This was a nightmare for anyone who would have been at one of these crossings, where the sheer numbers of trucks backed up, waiting for the new forms. This was especially true when perishable goods were involved, since the trucks dared not get out of the long line at the crossing, yet they had to shut down the vehicles during extended waiting periods while their organizations attempted to get the necessary forms delivered. The first representation of this is shown in Fig. 10.12 where the vehicles were delayed.

Those vehicles that were already equipped with a cellular phone and a facsimile machine experienced the least amount of delays. Hearing that they needed a special form, they immediately used the

Figure 10.12 The pre-fax scenario during the Gulf Crisis.

cellular phone to notify their dispatch office of the new requirement. From there the organization obtained the appropriate form, filled it out and faxed it directly into the cab of the truck. The trucks were then allowed to proceed with their assigned routes and delivered their goods. This is represented in Fig. 10.13 where the process worked quite efficiently.

This is also an application where medical records are needed to provide lifesaving techniques on the road for emergency personnel. Although this could also be done with the use of a data transmission, most patient records are already in written form, historical files are handwritten, and other forms of information may not be readily transmittable except for the use of facsimile.

Sales literature and brochures that are needed when things change are added applications for the use of a facsimile machine. The list could be extensive. To provide a facsimile machine in a vehicle there are options. The first is the use of an added module that will plug directly into the transceiver in the vehicle. This transceiver module will have two RJ-11 jacks on it for either a voice or a data/fax option. The first jack will be used to plug in the standard handset so that the cellular phone can still work as normal. The second jack will provide the necessary interface (the RJ-11 jack) so that a standard facsimile

Figure 10.13 The post-fax scenario allowed information to be shipped directly into the vehicle.

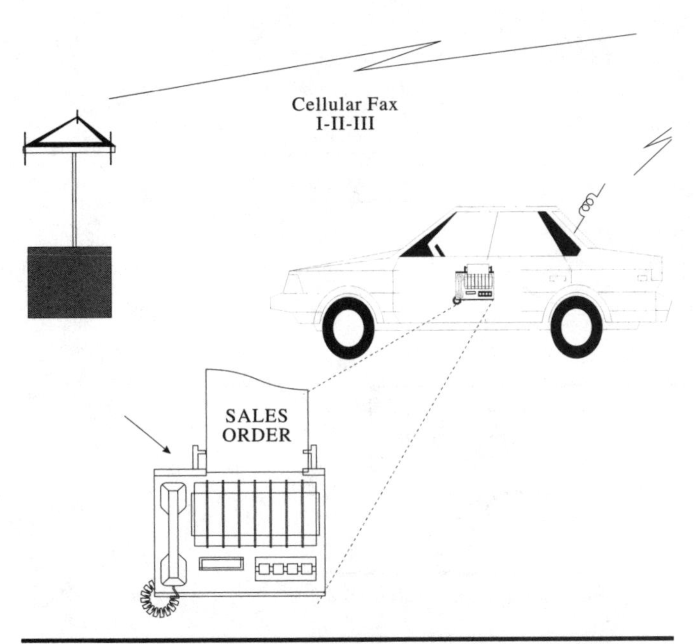

Figure 10.14 The use of a standard fax can be handled in a vehicle.

machine which has an analog modem built in can work. This may require the ability to generate the necessary dial tone (DTMF signaling) and the ringing voltages that are not consistent with the cellular sets. Therefore a card is installed in this box, with a cellular to telephone interface card such as those manufactured by Telular, Inc. This provides all the necessary signaling and tones for the standard fax to work in a vehicle.

Figure 10.14 is a representation of the fax machine in the vehicle supporting a standard CCITT Groups I-II and III. In Fig. 10.15 the actual box that will support the multiple interfaces is shown. This is the device that will support any standard teleset and data or fax modem capability as developed by Cellular Line Interface Corp. (CLI) in a small device that measures 1″ × 5″ × 4″. The problems that were addressed in the wireless data communications section above are not as prevalent with a standard facsimile machine. The fax uses a different data protocol where it uses a Huffman or Modified Read code. To transmit the information, the fax has a scanning capacity that scans on a line-by-line basis. As the scanner encounters the presence of a dot (pixel), it then transmits that dot down the circuit to the receiving end. However, if the carrier tone is lost, no data will be transmitted during that line timing sequence. The fax modem is more tolerant of

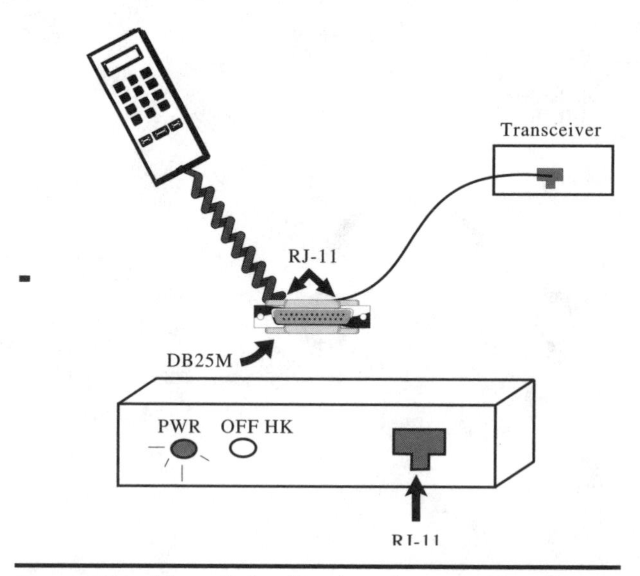

Figure 10.15 The cellular line interface provides the voice/fax jacks for vehicular use.

line hits and will typically wait for up to 400 ms for the carrier tone to return. In the event the carrier is interrupted, during a handoff for example, the fax will just not send any data for that period. This will not result in a terminated or dropped call, but the transmission of a blank line down the circuit. This is represented, although exaggerated, in Fig. 10.16 where an interruption of the carrier occurs and the resultant data received is shown in the drawing. Using this type of modem allows for a more effective form of data transmission or facsimile that is easy to use and permissive to the cellular network problems. The loss of data in a wireless facsimile transmission is minimal.

Another benefit of the fax machine capabilities is that in areas of poor transmission and reception, the fax machine has a fall-back and step-up process that will still allow data to traverse the network even if the reception is poor. The fax merely falls back to a slow speed for transmission and reception. However, in the fall-back mode, the machine continually scans the line quality and when this improves the machine will step up to a higher speed at 9600 bps. This delivers the best in robust communications capabilities for the wireless world.

This discussion is based on the true facsimile machine that would be used in a regular wired network and office environment. Future facsimile machines will be developed that are true machines but will use a cellular modem in lieu of the standard wired data modem in them. These machines will automatically use the wireless world with-

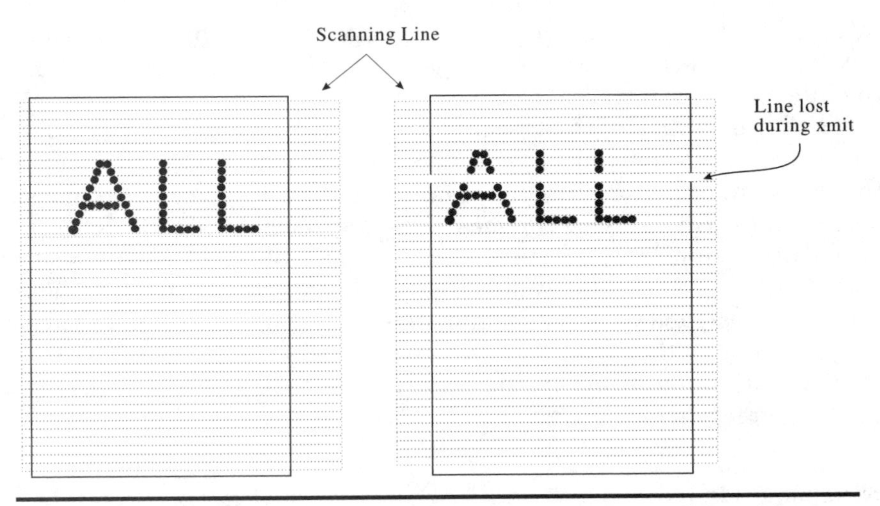

Figure 10.16 Interruptions on the circuit do not cause disconnects. A line of dots may be lost during handoff.

out the need of a special device as shown in Fig. 10.15. Instead it will be a stand alone device capable of sending and receiving fax messages directly across the cellular networks without the use of a cellular handset or other interface. This is a modification of the original concept, but it adds some dimension to the reality of the wireless communications world. These fax machines are designed around the need to move the device around, so they are ruggedized versions of the standard fax. They will support Group III CCITT fax transmission standards.

Braley Communication Systems is the major distributor of many of these devices that will serve in different capacities like the ruggedized fax called the "BrikFax." The BrikFax is the only fax machine that meets the automatic electronic specification for shock, vibration, and thermal range. This device is a 6-pound device with dimensions of 4 × 4 × 11 inches that is designed to be used remotely. It works with a 12 Volt DC power system and the car cellular phone (so it is not a standalone box). The BrikFax has special paper that is designed for temperature extremes and can support being dropped at up to 3 feet without any damage being sustained. A second device is the Command Post that combines data modem, voice communications, and BrikFax services in a briefcase mount. The single set is used in emergency systems where all three devices may be needed at one time or another. This device combines the cellular data and fax through a single modem interfaced to its own cellular phone link on any commercial or private network. Each Command Post includes internal batteries so that the unit can be operated by remote from any type of external power. A cellular interface is still required using the integrated systems.

Another deviation of this facsimile capability is the use of the PCMCIA card with a built in fax and data modem capability. Once again this is for an either/or situation. You can use only one or the other at a time; data will likely be the most frequent service. The PCMCIA, or an older version laptop with a standard ISA or EISA card, can be operated using a cellular connection to the rest of the world. These cards are designed to allow for the standard interface to the wireless world. Although these cards are expensive today, the pricing will drop to more palatable levels in the near future as standards and mass production levels pick up.

Wireless E-Mail

There can never be any doubt that the various abilities discussed thus far have viable uses within an organization. However, as users become more computer literate and the ability to use data terminal equipment (DTE) eases in as second nature to them, the use of the terminal will tend to displace the need for real time voice communications. In this regard then, the use of electronic messaging or E-mail is fast becoming more acceptable and preferable by this population. No longer do users want to run around with large computers, pagers and telephone sets, they prefer to receive their notifications and alerts via mail that can be stored until later retrieval or responded to immediately. The choices are many on how this could be deployed within an organization. The combination of mail and packetized data have merged together to form a whole new market, the wireless messaging industry. Wireless messaging is not new. It has been here for years. What is new is the availability and the options to users and business alike. Many longer time users who have been in the data-processing industry have recognized that IBM for several years equipped the field service technical personnel with a device that was fondly referred to as the "brick." This was a wireless messaging service that was set up for use exclusively by the IBM personnel.

In 1990 a joint venture between IBM and Motorola named the Advanced Radio Data Information Service (ARDIS) was formed to provide a nationwide wireless data network. Access would be through the ARDIS network for subscribers and field service organizations. The network was already deployed to over 400 cities in the 50 states for U.S. distribution. The network operates on a single duplex channel pair (25 KHz channels spaced at 45 MHz apart in the 800 MHz frequency band). ARDIS is a series of overlapping cells to provide extensive coverages within buildings in major metropolitan areas and even in less populated locations. The overlapping nature of the cells increases the probability of capturing a signal. The network's overlapping base station controllers sample the signals coming in across the network from an ARDIS unit, then determines which transmitter to use. Typically three separate controllers will be able to sense and control the data, but one of the three will be receiving the signal the strongest. It is this strongest receiver that will decide to take control of this particular transmission. The network runs at 4800 Bps on the radio frequency (RF) however, approximately 60 percent of the data is allocated to overhead netting a true data throughput of approximately 2 Kbps. Upgrades to the network being deployed around the country will increase the overall network capacity to approximately 19.2 Kbps and yield approximately 8 Kbps of net user data. ARDIS is a packet switched network, built on a proprietary protocol therefore creating a closed environment for internetworking between other networks. The cost of use is on a packet basis which a typical packet consists of 256 Bytes of information and costs between $.05–$.10 each. In Fig. 10.17 the overlapping receiver platform is shown. The three receiver triangulation in this figure indicates that in tall buildings where concrete and steel could be a problem for the wireless communication, the problem can be overcome by a controller sampling technique.

MOBITEX is the brainchild of L. M. Ericsson and is a registered trademark of Swedish Telecom for the formation of a public packet

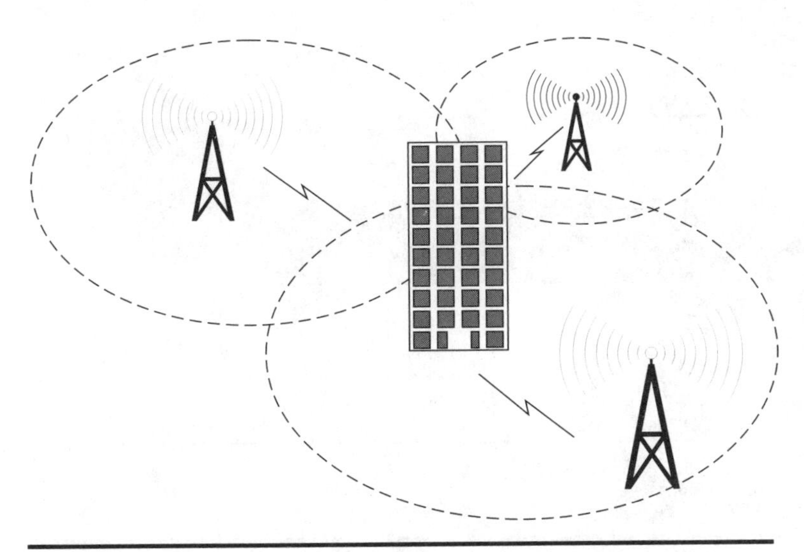

Figure 10.17 The ARDIS concept uses overlapping receivers to select the best reception.

switched wireless information network for electronic messaging, any form of data transport, and paging services. RAM Mobile Data (a joint venture of RAM Broadcasting Corp. and Bell South Enterprises, Inc.) is the responsible agent in the United States to construct and operate this network. RAM has from 10 to 30 channels in each of the 100 top MSAs in the country. The platform is based on an open architecture, unlike most of the others. RAM currently operates in over 6300 cities and towns. In the present configuration this network operates as a packet data communications network. RAM uses the public networking technology known as Mobitex. Mobitex networks are in use in the United States, United Kingdom, Canada, Sweden, Norway, England, Chile, Australia, Netherlands, and France. It is built to offer seamless automatic roaming so that subscribers are addressable as users, instead of places. The network will automatically find the user anywhere within the coverage area. All the user has to do is turn the unit on. The data packets are designed around a standard packet that is shown in Fig. 10.18 which are called MPAK packets. This packet is similar to an X.25 packet with approximately 24 bytes of overhead and up to 512 bytes of user data in the MPAK. The cost of this connection is based on variable information. However the initial costs of $25.00 per month per device and a cost per packet ($.05 per 100 byte file or $.125 per 512 byte file) is the basis of their network systems. The communicating device, called Mobidem, is a one pound device powered by batteries and has firmware with a hard coded ID number

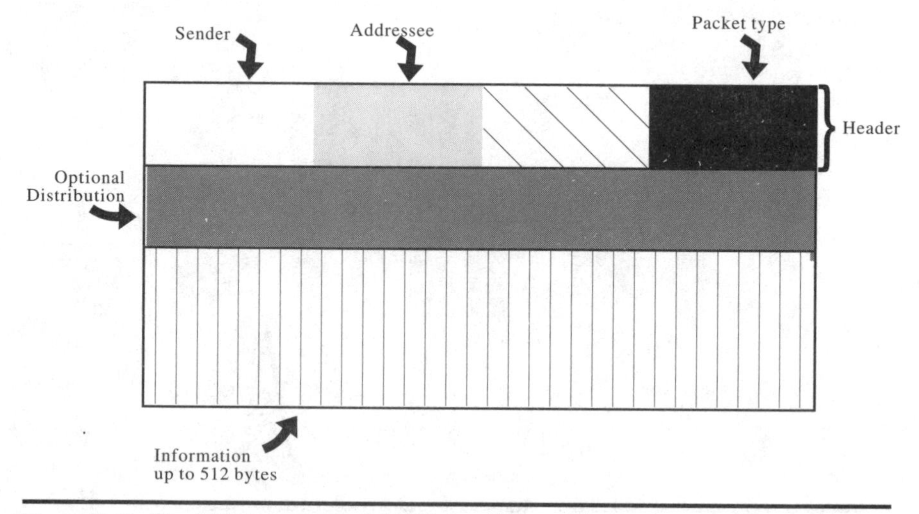

Figure 10.18 The Mobitex packet (MPAK) handles up to 512 bytes of information in packet form resembling an X.25 packet.

that transmits and receives data at slightly less than the 1 GHz range. Mobidem connects to a radio network of radio towers operated all around the United States.

Each Mobidem identifies itself to the nearest tower and upon successful establishment of a connection the unit broadcasts whatever information it has in an 8 KB buffer. The data rate is up to 8 Kbps depending on the quality and clarity of the connection from the handheld device to the tower. The connection is a function of the distance from the tower, obstacles in the path such as concrete and steel, and any other path fade conditions. If a user happens to be out of range of the system, the device will store the messages in a queue and wait until it enters into range of a tower. Once the user is in range of the system, it will broadcast its stored information. Ericsson came up with various arrangements to configure and sell access into the Mobitex network, including a Mobidem (modem) with a palm-top computer such as HP 95LX, a cable connector to link the devices together and the necessary software to run the messaging system. Linking this package with software from RadioMail Corp., a PCMCIA (or flash) card capability is provided that will allow a user to use the messaging software of the system. The RadioMail software allows access to various other networks through Gateway software for MCI mail, Internet access and CompuServe. Several vendors are now licensing and marketing the Mobidem, to include the recent introduction of the Intel Wireless Modem. Moreover, companies like Microsoft, Lotus Development Corp., DEC, AT&T, Easy Link, Da Vinci, and others have endorsed Mobitex communications capabilities. The company (Ericsson) registers the Mobidems with RAM Mobile Data and the logging of the device with the FCC. RadioMail bills for the usage based on a packet basis. Initial costs for this service are in the $1.00 per page of fax plus an $89.00 basic monthly fee (this includes the first 100 messages as part of the monthly fee). The Mobitex system stands to become the standard for wireless data messaging due to its wide acceptance and vendor support, plus the openness of the architecture.

Telepartner International has developed its Mobi/3270* for wireless data transmission into an IBM mainframe world *ala* the 3270 computing arena. This uses several different components that will allow remote access for mobile users. The pieces combine the following: Packet/Main software that runs at the host end to allow for the VTAM program to relay the information between the application and the remote PC. The other piece is a software solution at the PC or ter-

*Mobi/3270 is a registered trademark of Telepartner, International.

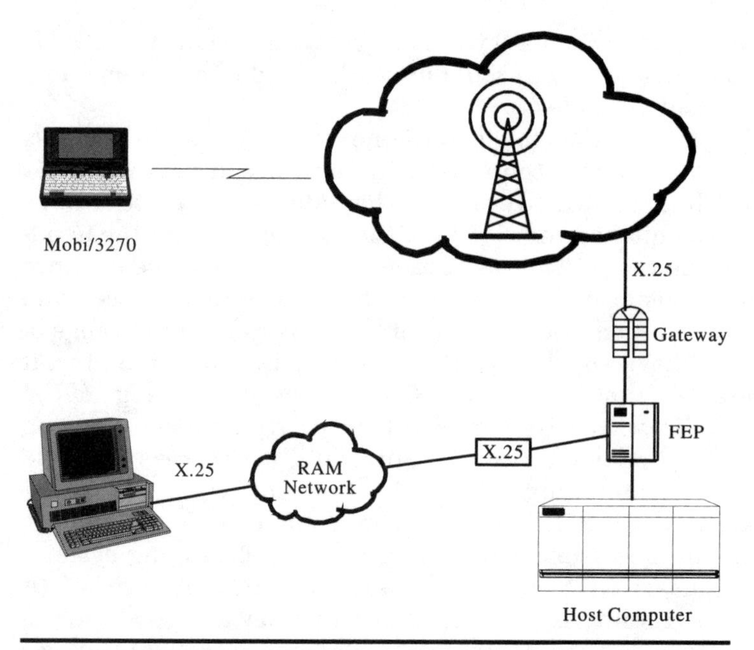

Figure 10.19 Telepartner international's access to a 3270 world from a wireless messaging system.

minal level of which there are many versions. This will use an X.25 gateway into the RAM Mobile Data network or a direct Mobitex interconnection with X.25 capabilities from the Mobitex switch. In Fig. 10.19 this is shown as a representation of the Telepartner service.

Several other players will emerge in the network now that standards have been decided upon. This will lead to new competitors and offerings along the way. Prices will become more competitive and other services will emerge as add-on capabilities for the network user. These are just the entrants into the burgeoning market for the e-mail and packetized data services. As users demand new forms of mobile connectivity, the developers and vendors will find new services to meet the demand.

Chapter

11

Use of Wireless in Disaster Recovery Situations

One of the most appropriate uses for wireless communications is during a disaster. A wireless connection offers flexibility and mobility unsurpassed in the wired world. To be sure, many organizations have already found very specific applications to have a wireless capability available, initially for a catastrophic loss where recovery is easily established. However, many of these organizations who have set a wireless technique in place as a recovery tool, have since migrated to using the service as a primary means of communicating too! This is not an indication of a move away from the use of wired communications, but more of a complementary product or service that can aid in quick moves, adds, or changes to the existing wired base. No single solution will fit all scenarios. Therefore, continued experimentation is a must for those who will attempt to move in this direction. Each telecommunications department, MIS department, or LAN administration should have some form of research and development (R&D) budget to experiment, test, and understand the ramifications of using a wireless supplemental or complementary technology. First, the use of cellular communications with the 30 KHz channel capacity can offer some strong arguments for the recovery of a voice dial-up network service. Through the use of a cellular link, the telecommunications department could provide a diverse routing capability to the outside world in the event of a cable cut or failure. Along with the single channel capacity of a cellular connection, microwave radio transmission has also been around for years as an alternate link to the outside world. Many organizations have used a microwave shot to provide large amounts of bandwidth in the T1 and T3* range to their

*T1 is a digital transmission system where 24 channels of 64 Kbps are multiplexed together into a 1.544 Mbps data stream, T3 is a digital transmission system where 672 channels of 64 Kbps are multiplexed together into a 44.736 Mbps data stream.

central offices or as a bypass technique to the interexchange carrier's points of presence (POPs). The application for this was emphasized when a fire occurred in Hinsdale, Illinois on May 8, 1988. The loss of a central office, one that also served as a gateway for cellular communications and for fiber links, put thousands of users out of contact with their customers and subsidiaries for extended periods of time. A solution for many was to use a microwave link into the networks of AT&T, MCI, and others. However, this must be planned and coordinated in advance where possible because of licensing requirements, clear-path considerations, and frequency coordination issues (see Chap. 2).

An alternative to microwave for long-haul communications or for the broadcast capabilities to many remote sites is satellite communications capability. The use of satellite has been a secondary technique in the past, but due to the broadcast from one to many simultaneously, it has also been used as a primary service. Satellite was also covered in Chap. 2 and requires the use of a commercial satellite carrier's network. This again is something that should be pre-coordinated before jumping into the service as a thing to do.

In the event of the loss of a building or a telephone system service, the wireless PBX can provide quick connectivity for multiple users to a PBX-like service. These are now being provided by vendors as a "recovery system on wheels" that can be rolled into a site within a specified period of time (24 or 48 hours). The wireless PBX is only a short-term alternative to the real system, but it gets the user back up and running with limited wiring and configuration constraints. Once again, the use of a quick, untethered communications process can bring more to the table than strictly the sizzle of a new "latest and greatest" technique. The wireless PBX is a possible alternative to issuing a group of users individual cellular telephones that will be a far more expensive proposition.

From the PBX arena, in the event of a fire or flood in a building, a lot of rewiring for a LAN connectivity will be required. However, as the technology evolves to bandwidth up to 20 Mbps, the wireless LAN can be a quick fix to bring up the necessary connection between users' computing platforms in a relatively short time. Many of the wireless LAN systems discussed in Chap. 9 are easily installed within the confines of a departmental situation. Personally the author has set up several wireless LANs with a mix of 8–20 nodes in less than one hour. This is obviously more expensive in the initial cost stages than physically pulling wires, but the need for a short-term connection or the need for temporary connections dictates that this could be the alternative, despite the cost implications.

Other such situations have existed in the past where the wireless technologies have paid off significant dividends in the communications capabilities. Some of the disasters where the wireless communications services of E-mail, cellular data communications, and other services is summarized in Table 11.1. Wireless LANs have been used in many situations (that will be discussed) after a natural disaster such as Hurricane Andrew in the southeastern portion of the United States, after the World Trade Center bombing in New York City in early 1993, and as a solution to a different crisis, The Gulf Storm crisis, which may or may not be considered a disaster to some. Regardless of the classification of what constitutes a disaster, the issue is that wireless LANs were installed quickly and efficiently to solve a need.

Linking LANs and PBXs together after a disaster can also be accomplished through the infrared laser connection discussed in Chap. 3. This section dealt with the short-haul communications connectivity through the use of unlicensed technology that can be a quick fix to a problem and a work around to the stringent requirements of a licensed radio-based system. To this end, the use of the infrared sys-

TABLE 11.1 A List of Where Some of the Disasters Struck and the Supplemental Services Used

Type of disaster	Applications used
Fire in Hinsdale, Illinois	Loss of a central office (CO) in a telco caused loss of total communications. Therefore all applications were used—microwave, satellite, cellular, wireless E-mail, and infrared.
Hurricane Hugo in the southeastern United States	Cellular, microwave and wireless data, E-mail applications all were used extensively
Earthquakes in San Francisco, California	Cellular, microwave, satellite and wireless LANs, wireless E-mail and data communications.
Floods in Chicago, Illinois	Wireless E-mail, cellular, and wireless data communications
Riots in Los Angeles, California	Wireless E-mail, cellular, and wireless data/fax
Hurricane Andrew in Florida	Cellular, wireless E-mail, wireless data/fax, satellite, microwave, wireless LANs, wireless PBX, infrared.
Operation Desert Storm in the Persian Gulf	Wireless E-mail, wireless LANs, wireless PBX, satellite, microwave, cellular, other.
World Trade Center bombing in NYC	Microwave, satellite, cellular, wireless E-mail, data, fax, and LANs

tems will be dealt with as a disaster recovery mode for the local connection or as an access means to the longer-haul communications capabilities. A standard set of interfaces will allow this technology to be set up to serve most needs within an organization for the mobile needs of the recovery process. Be it LAN or PBX connections, this is still an inexpensive solution to have on hand for the dynamic setup of a communications bandwidth.

Lastly, as the whole picture comes together the mix and match of these technologies is possible. As long as standards are adhered to the use of satellite, microwave, infrared, and cellular communications can all be managed and massaged into a single homogeneous network if necessary. No single selection of a technology solves all problems, but no single solution should limit the disaster recovery efforts of an organization. An understanding of the strengths and limitations of each of these techniques should now have been achieved through the previous chapters. Applying what has been discussed, the systems can all be interlinked if need be to solve the needs of the organization.

Using Microwave for Disaster Recovery

The first of the many techniques that should be considered is the use of large amounts of bandwidth to connect to the outside world via a microwave system. In many cases the organization that is concerned about a phenomenon called "backhoe fade" or the cutting of the wires entering a customer's premises, will be looking for an alternative access to the building from the outside world. Many alternatives work here including the following options:

- The use of diverse cables from the local telephone company. In this scenario however, many customers and telephone companies have been stung badly. The sting came from a misconception about what constitutes diversity. In the telephone company engineering groups the diversity usually dealt with cables. In the cable arena the diversity could be in the form of one of the items listed in Table 11.2.

  In Fig. 11.1 these are summarized to be more specific in graphical representation.

- The use of different wire centers to serve a customer location is a newer technique used by the telephone companies after the problems of fires and floods were experienced. This still dealt with the use of cable facilities to two different COs rather than two routes to

TABLE 11.2 The Forms of Cable Diversity

Forms of diversity
1. *Bundle diversity,* in which the diverse route between the customer and the telephone company building is in different bundles of wire. However, the bundles may, and usually are, in the same conduit right of way. Therefore, if a cable cut occurs from a backhoe the likelihood of both bundles being cut is very high. This does not provide much assurance of a recovery technique if the primary and alternate cable can be cut in the same fell swoop of the backhoe.
2. *Count diversity,* in which the primary circuits entering the customer's location or the telephone company building are in the same bundle of wires, for example 600 pair cable. However, the diversity comes from the telephone company splitting 50 percent of the wires in count 1–200 and the other 50 percent of the wires in count 400–600. This is obviously a major problem, since all the wires are in the same bundle, just on different bundles in the overall cable. One incident will most likely put this out of service.
3. *Sheath diversity,* in which the sheaths of wires in a much larger bundle; 3000 pair cable, for example, are separated and the diversity is applied to different sheaths. The 3000 pair cable is made up of 600 pair subcables similar to the count diversity above. The telephone company would therefore put 50 percent of the circuits in one sheath of 600 wire pairs, and the other 50 percent of the circuits in a second 600 pair sheath. Both of the sheaths are contained in the larger cable structure so a problem is very likely to spread between the multiple sheaths. This is particularly true with a fire or flood, as well as a cable cut.
4. *Route diversity* is probably the only true form of diversity that could be provided with any sense of security. In this arrangement two separate paths are used between the customer's premises and the local telephone company's location (the C.O.*). The unfortunate problem with route diversity is that there is still likely to be a single point of failure where the wires will come together, such as at the cable vault in a building or the entrance to the C.O. The whole concept of route diversity is that two separate paths are used and that no place along either route will the cables come across a single point. This is very expensive to accomplish with estimates in the past ranging from $80,000 to $250,000 for this service. Unfortunately, the risk here is that the two separate routes still lead to the same building and CO. Therefore, if a major disaster strikes in either the CO or the customer's location, the two distinct routes are for naught, since the building is inaccessible.

*C.O. stands for the central office or end-office that provides the dial tone to the customer location from the telco facility.

the same CO. This is shown in Fig. 11.2. Although this is an improvement the problem still exists through a single entrance point at the customer end such as a cable vault. Further the loss of access to the building rules both routes unusable. This is an expensive technique to run cable in two different directions to two different COs, only to realize that a single problem could still render this useless.

- The use of different carriers to access the building. The use of the competitive access providers (CAP) fiber optic connections through

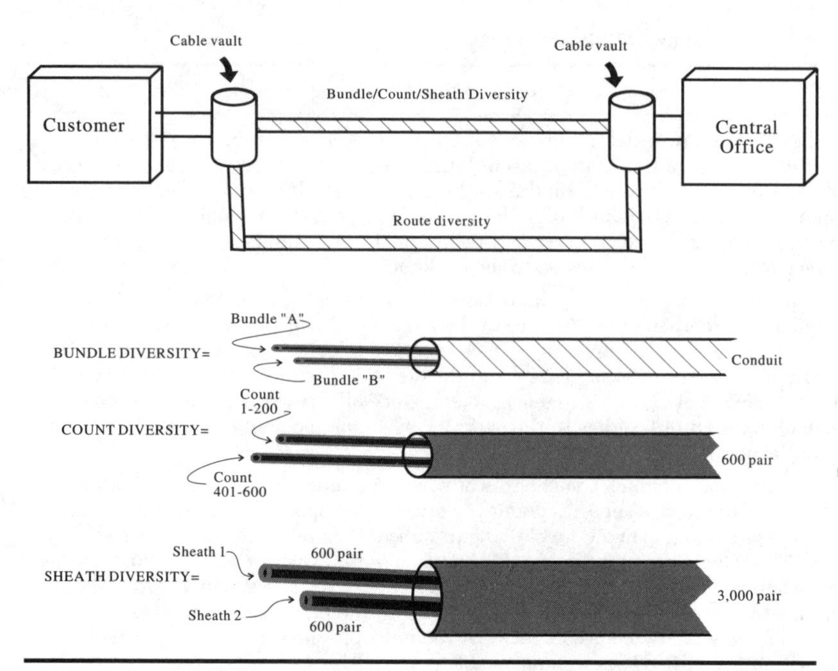

Figure 11.1 Diversity takes many forms in a TELCO world. Route diversity is preferred but more expensive.

a ring topology from the CO to the customer premises and the use of the telephone company wires as an alternative is a more widely used technique these days. However, the same problem exists if the loss of the building occurs. All of the access points to the customer locations are useless and nothing has been gained. Further,

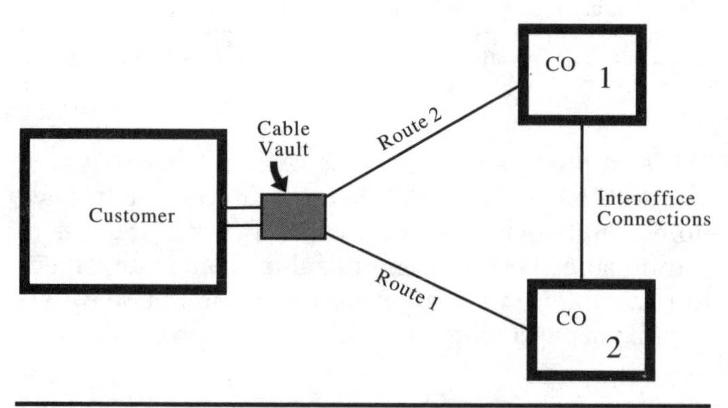

Figure 11.2 Running different COs doesn't totally solve the problem. A single point of failure exists at the customer end.

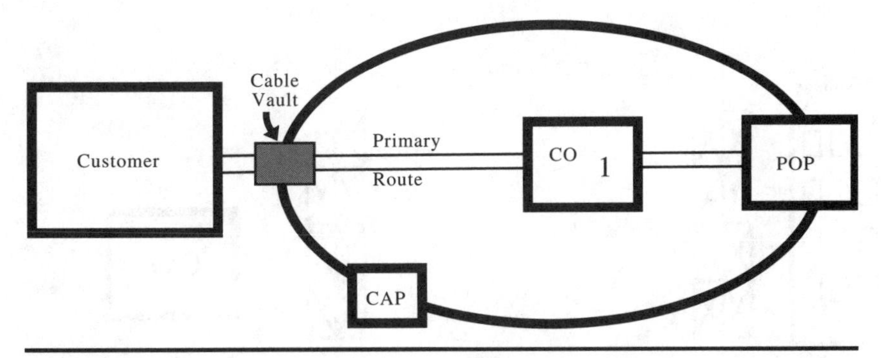

Figure 11.3 The use of a CAP service may offer some robustness to disaster recovery prevention. However the CAP may still pass through a single point of failure negating the diversity.

with the use of co-location services* many of the CAPs are using the same entrance points and in some cases the same right-of-way into the customer location thereby exposing both sets of service to a single failure on the cable route. This is shown in Fig. 11.3.

- Finally, an alternative technology such as microwave can be a means of getting access out of the building to the central office, as shown in Fig. 11.4. The microwave system can be used to support the capacities shown in Table 11.3. The microwave system in this case is provided by the local telephone company as a disaster recovery tool and therefore is not an expensive proposition for the provider or the customer. In this particular scenario, the use of the microwave is for a second means of accessing the same CO.

  However, the microwave can also be used to access a second wire center, much the same as a wired facility was provided in the past. Therefore, if the primary CO fails, the microwave as the secondary route can provide some services (possibly all) from the second CO. This is shown in Fig. 11.5, in which the microwave system serves the secondary route and the cable facilities serve the primary route.

  A third alternative to this scenario is the bypass of the CO where the long distance carriers will offer to co-locate the microwave as a means of accessing the long distance network if the CO fails as shown in Fig. 11.6. This is a newer approach. The primary circuits from the CO are serviced by the wire and the secondary (or recov-

*Co-location services is the ruling by the FCC and the courts that the competitive access carriers are allowed to locate their equipment in the telephone company CO to competitively offer dial tone or other leased line services competing with the LEC.

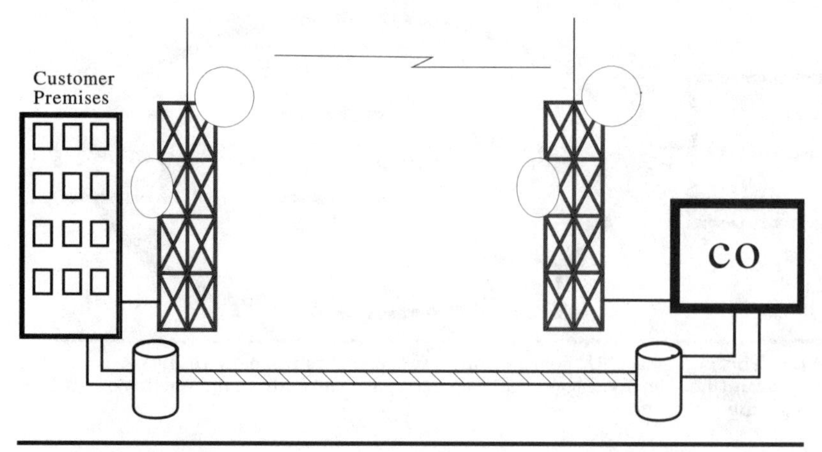

Figure 11.4 Microwave provides a diverse route to the CO from the customer's premises.

ery services) are served as the bypass route. The benefit of this approach is only if the CO or the cable between the customer premises and the CO fails. This is still not a solution if the customer building is the one suffering through the disaster.

Therefore, a portable microwave system or a temporary system can be used if the end-user facility fails and the customer must be relocated to a new building. This will likely be quicker to use in the event of a disaster than waiting for wires to be run to the new or temporary location. Clearly, the local telephone companies will make their best efforts to get a customer up and running as quickly as possible. However, if this is not an isolated building problem but something affecting many buildings or a whole area, the timing could become far more critical. Thus, the use of a temporary shot would clearly fit the need. In a disastrous situation special provisions can be made to get a system up and running quickly, working around the problems of long lead times for licensing etc. Further, many of the carriers have the licenses issued and merely have to conduct the necessary path and interference studies to get the ser-

TABLE 11.3 The Typical Capacities Provided on the Various Microwave Bands

Frequency band	Typical capacities provided
2–6 GHz	8 T3
10–12 GHz	4 T3
18–23 GHz	8 T1 or 1 T3

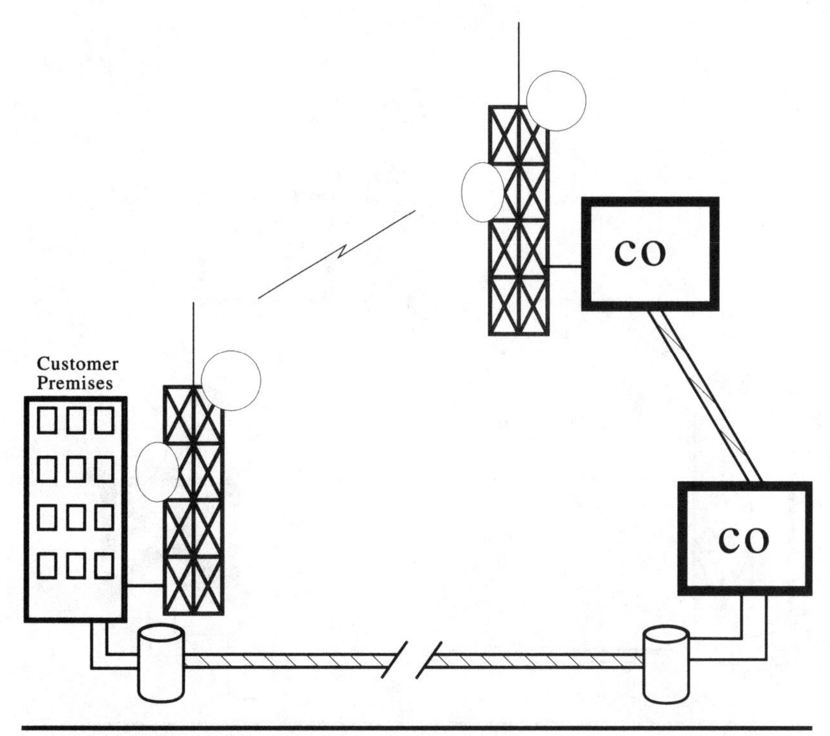

Figure 11.5 Microwave to the CO can add new dimensions to the prevention of outages.

vice working. This will prove to be a valuable asset if the need should ever arise. Another option is the use of a portable 18–23 GHz system, that can be transported to an area that has suffered through a disaster. The use of a van mount or merely a carry around capability could be sufficient to get enough connectivity up and running for the short term. In Fig. 11.7 is a van-mounted system that can be rolled up to the site of a disaster and commissioned into service quickly.

Use of Satellite Communications for Disaster Recovery

A number of organizations have installed a satellite capability as an alternate route capability for the longer-haul communications channels they currently use. In particular, the use of high-speed digital trunking via satellite has been fairly commonplace for data-processing links. The use of a land line across country has been migrated to high-speed fiber optics circuits. These are considerably more reliable

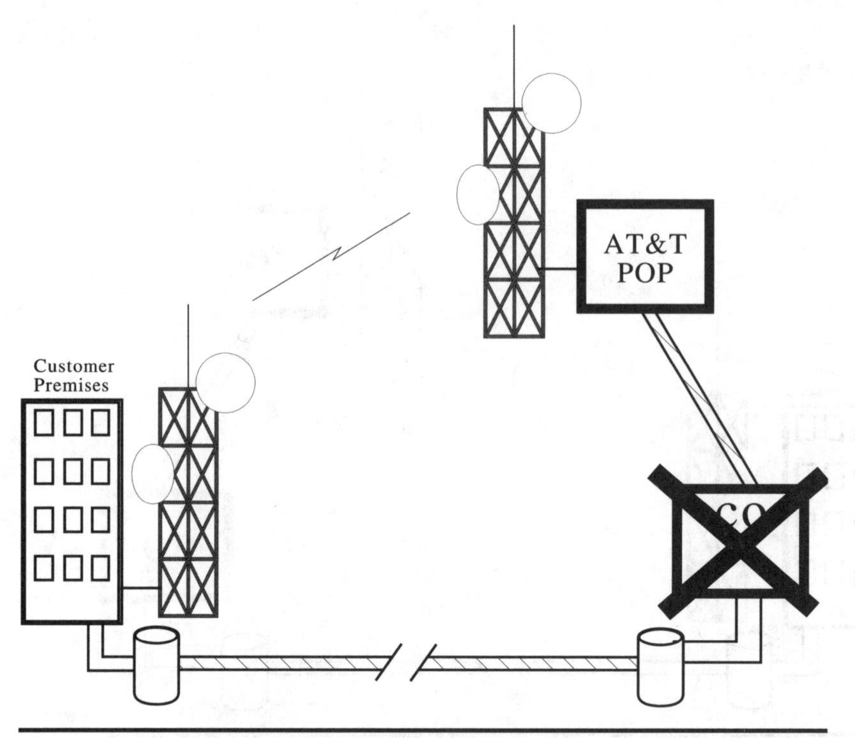

Figure 11.6 Bypassing the CO to the POP allows access to the long distance networks even if cables are cut or the CO fails.

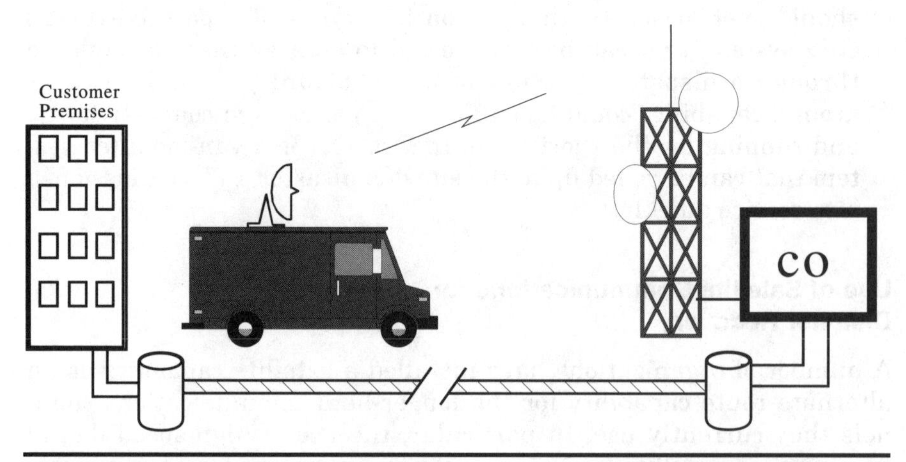

Figure 11.7 A van-mounted portable microwave system can be used as an alternative in an emergency.

than the older copper-based (twisted pair wires) network techniques of the past. The bit error rate for fiber has also improved the transport of high-speed data between sites. This is the value of the fiber channel. However, all too often these fiber links are installed in the same rights-of-way that the copper wires are installed. The wires and the fibers are side by side in the same direct buried conduit runs. Thus, when a cable cut occurs, the copper and the fiber links are severed at the same time. Suffice it to say that the fiber carries so much more channels than the twisted pairs of copper, that when a cable cut occurs, more channels are affected concurrently. Thus, users had to look at the alternatives.

- The first is to build a ring topology with the fiber loops. The carriers have been deploying fiber rings to overcome these problems. But, this is an expense for the carrier that must therefore be passed on to the consumer. Consequently the use of a ring topology, although considerably less than in the past, have met with only moderate acceptance by the end user. A minor problem also occurs when the carriers have to use rights-of-way for the fiber, there always seems to be a single point where these fibers all cross in the same path. Consequently, when a disaster strikes, it will be at this point of convergence.
- To overcome this problem, many users have opted to look at the wireless side of the business as a second alternative. The use of an older technology, the satellite link can offer large amounts of bandwidth and be less restrictive. As a broadcast technology, the satellite links have been used as a recovery technique for cable cuts and for relocation needs. In Fig. 11.8 the use of a very small aperture terminal (VSAT) has made significant inroads toward backing up critical data, voice, LAN and video circuits. The use of the VSAT allows an organization to diversify the risks of outages and disasters, since the signal is broadcast from one location to one or many remote sites. The VSAT has become far less expensive for the terminal equipment. Further, this is supported in the knowledge that in 1989 there were approximately 20–25,000 VSAT terminals in use. In 1993 the number has increased 10–20 fold depending on how they are counted.
- Portability of the VSAT application is possible. Vendors have roll around satellite dishes that can be rolled up to a customer location. This is shown in Fig. 11.9 where a satellite dish supporting up to a T1, or multiples up to a T3 if capacity is available from the carrier, can be delivered quickly to a customer location for the institution of a recovery capability.

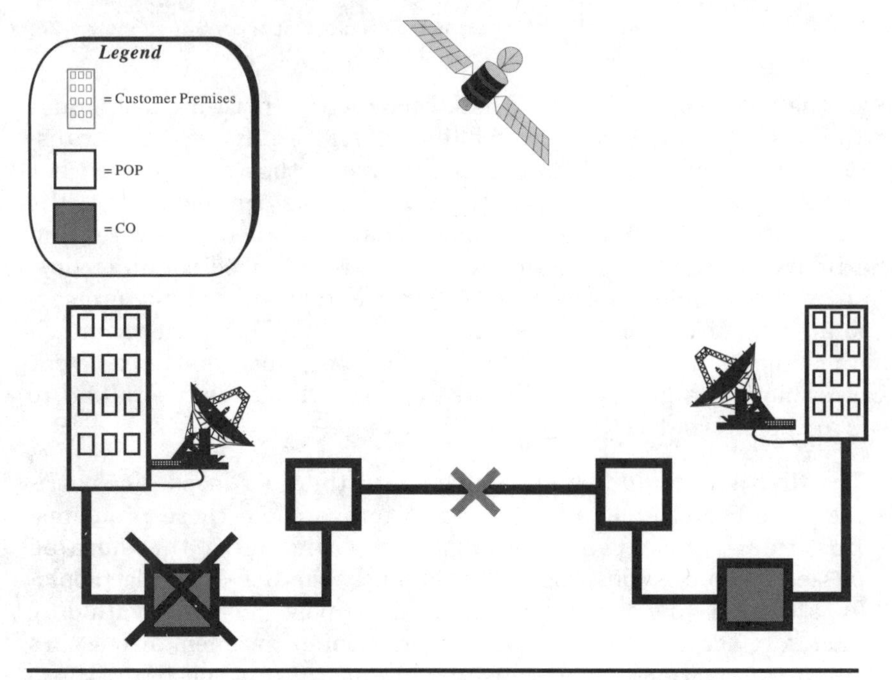

Figure 11.8 Critical voice and data can be supported on VSAT terminals in the event of a failure anywhere along the links.

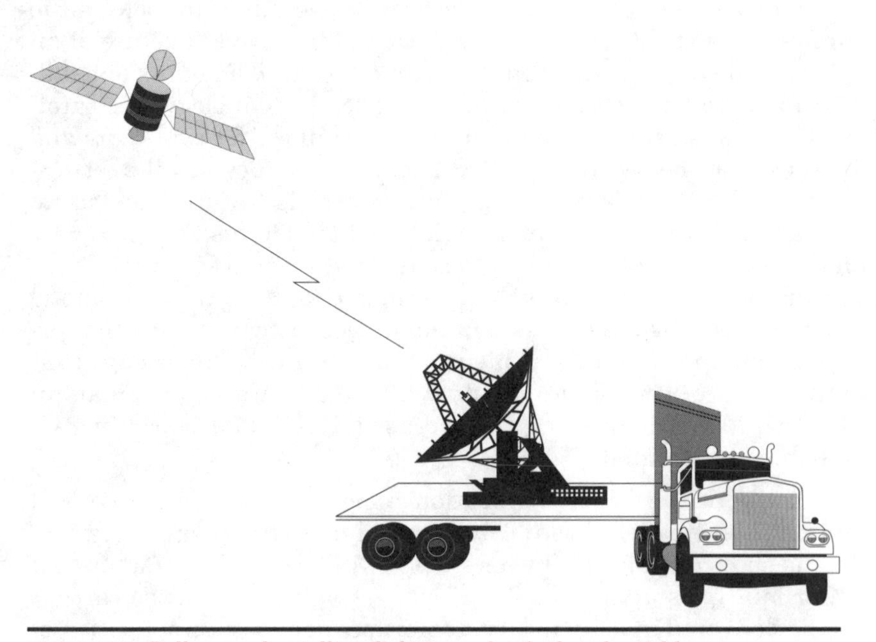

Figure 11.9 Roll-around satellite dishes can be deployed quickly to a customer location.

- An even easier method of getting service quickly is the use of the suitcase model satellite terminal. This is a single or dual suitcase with all the electronics and dish to support lower-end needs such as a voice, data, and fax service for an individual. Remember the Gulf crisis, when the news media all had to have the ability to establish links back to their respective countries and the service availability was limited. As a result, these organizations shipped the VSAT terminals out to the remote areas for quick call set up for voice especially, but also for data and freeze frame video. These are not inexpensive, but they could fit the need.

These suitcase or trailer mounts have been used by the Federal Emergency Management Agency (FEMA) in responding to the needs of communities after a natural disaster. The FEMA personnel set up portable satellite dishes and communicate back to their headquarters or to a central site with status reports, pictures of devastation and statistical information that is managed as part of the recovery effort. Business, government and other agencies have all recognized the need for the wireless connection. The particular use of this satellite communications capability has been effective in the disasters listed in Table 11.4. The

TABLE 11.4 A Sampling of Where Wireless Satellite Communications Were Used

Disaster scenario	Services
Hurricane Hugo	Voice and data on the Caribbean islands as the only source source of telecommunications available.
Hurricane Andrew	Voice, data and LAN traffic all carried for government and and businesses alike. The FEMA personnel used this as one of their primary connectivity solutions.
Earthquakes in San Francisco	Many businesses realizing that their telecommunications services would be affected in the wired world effectively used the satellite communications to reestablish connectivity back to the mid-west and east. The use of TDMA and DAMA* were used extensively.
Fire in Hinsdale central office	Use of satellite for data and LAN traffic was used heavily as as users scurried around trying to find a means of communicating with the rest of the world. The satellite channels-satisfied many of the long haul communications needs.
Flooding Mid-West	The cables and many of the central offices were under water in a raging river that surged over its banks. Therefore many of the traditional wired network facilities were useless. The use of satellite and microwave links help to get voice, data and LAN traffic running smoothly again.

*Demand assignment multiple access (DAMA) is a usage-oriented service. The customers may not have fixed amounts of bandwidth allocated for their use, but on a demand basis, the customers can use as much available resources as possible.

above sampling of disaster situations dealt with the larger, more obvious natural disasters. However, many other disasters occur on a daily basis, affecting only a small number of organizations. These are not as publicized. Therefore, the use of the technology goes unnoticed.

Using Cellular Communications for Disasters

Clearly no other service has emerged as a logical evolution to complement or supplement the use of the wired communication networks than cellular communications. The world still transmits a lot of information verbally, with voice still reigning as the primary service being used. There is no doubt that data and LAN traffic are escalating and will surpass the use of the network for voice. This is still in the future, so the comparison of cellular communications is a must to replace voice dial-up links. The cellular networks have emerged in disaster recovery modes in various forms. In some cases the disaster is of such monumental proportions that the entire telephony network is down for short to extended periods of time. Although this is being dealt with by the local telephone companies, no plan can be a perfect one.

During the hurricane scenarios of Hugo and Andrew, the devastation to massive areas caused severe outages across the wired network. Many times the problem was not that the network was down per se, it was that the wires connecting the user to the supplier were either downed or were flooded. Therefore the connection was impaired significantly. The central office worked but the end user was still out of service. To overcome this problem, users were able to establish a voice connection quickly if they had a cellular phone. In the disaster recovery mode, the users are now creating lists of who in the organization has cellular phones that are company or organization owned that could be reallocated quickly to get the business of the business up and running quickly. To use the cellular network offers no major risk (other than those discussed in Chap. 7) but the expense is an issue that cannot be taken lightly. The choice is if the organization needs to be communicating quickly, at any cost; or if the organization can wait. The cellular networks have never experienced any significant downtime since their inception. However isolated downtime has been experienced. To take this a step further the cellular carriers have never enacted any of the rules of recovery that are typically applied by the local telephone companies under the TPRS*

*The Telecommunications Priority Restoral Services is a logical restoral of services allowed for the telephone companies to reestablish service to various organizations.

TABLE 11.5 These Sequences Are Followed by Telephone Companies Under the TPRS

Summary of steps used in a disaster scenario
1. Immediately after a disaster the telephone company shuts down all access to the central office by denying all dial tone to new calls.
2. The telephone company personnel check the network and connectivity of the central office for any corruption or other possible outages or interruptions.
3. The service is then restored in an orderly fashion by allowing users in logical groups to access the dial tone.
4. National defense and security organizations are the first to be restored. Following the national security the next group to gain access is emergency organizations such as hospitals, and fire and police departments (911) etc.
5. Only after the logical services are restored for the life and limb organizations, does the telephone company restore pay phones then users (i.e., business and residential services).

guidelines. An example of what is being led to here is summarized in Table 11.5, highlighting the steps of recovering a telephone network under emergency conditions.

Although this sequence above may not carry any bearing on the typical business because the sequence happens fairly fast, many organizations are concerned over even the shortest interruption to their business communications. The interruption to many organizations may lead to hours of downtime, due to the need of resetting everything, then bringing up terminals, sessions etc., in a logical sequence. When considering that the cost per hour of downtime for many organizations could be in the hundreds of thousands to the millions of dollars of productivity and revenue generation, it becomes clear that any form of communications will be an asset.

To prevent downtime of the wired world, the use of a wireless backup is therefore imperative. The cellular network allows the quick call setup on a dial-up basis to get things back in order quickly. Now that the wireless cellular network can also be used for the transmission of data at up to 19.2 Kbps and for the transport of facsimile messages and E-mail applications, its use is far more robust and critical. Since the TPRS has never been used in a cellular world, the issues of call and load shedding are not as demanding. As standards are emerging, the use of the cellular network offers a myriad of opportunities in a disaster recovery mode.

Lumping PCS/PCN with the use of cellular, the future offers the ability of organizations to quickly keep business running with limited interruptions. Where the network can connect to a user at any location at any time, the issue of downtime at the business location goes

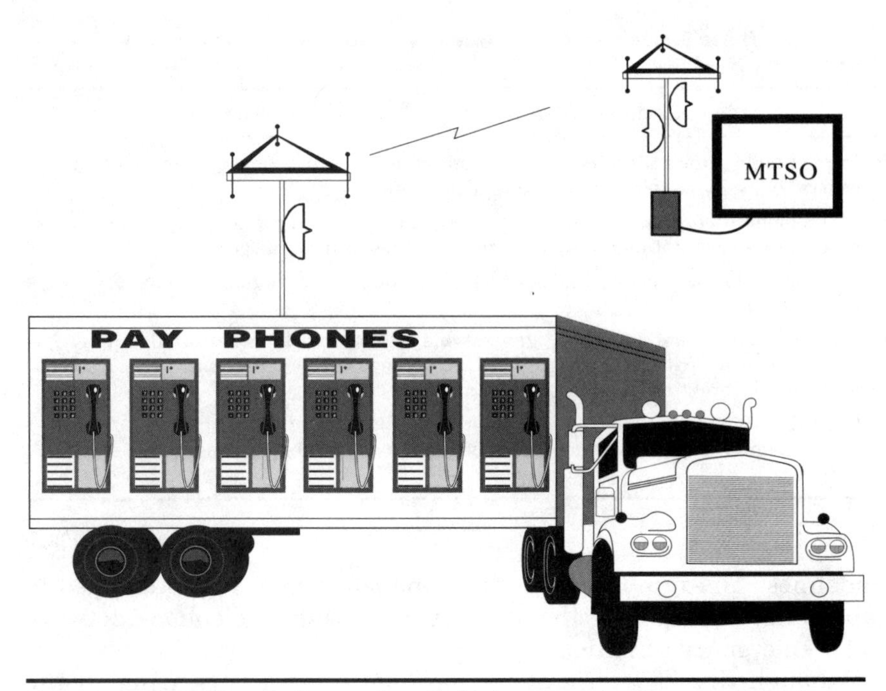

Figure 11.10 Portable cell-sites can be rolled into an area to serve payphones in an area struck by disasters.

away. It would only be the network that would be a concern. However, since there will be so many players and providers in this industry then options will be far more readily available to use a dual connection or a dial-up on-demand service. These capacities for the transport of the business information are just the beginning of the future uses of these technologies.

Companies have portable roll up cell sites that can be brought to the site of a disaster. In Fig. 11.10 a portable cell site can be rolled up to an area where a natural disaster has struck. Several wireless pay phones can be externally mounted on the side of an 18-wheel trailer which houses the cell site. The users therefore have to go to the closest telepoint (i.e., the trailer) and make calls. This could prove beneficial for the masses, but what of an isolated building situation where the portable cell site on a trailer may not fit? The answer is in a captive portable cell site where a site can be mounted in a customer's building for use in case of a loss of cables or a central office. In Fig. 11.11 the use of a captive cell site (purchased by a company) can be connected to the existing telephone services. This device is mounted in the top of a building and has the capability of up to 90 channels of capacity using today's analog cellular communications technology.

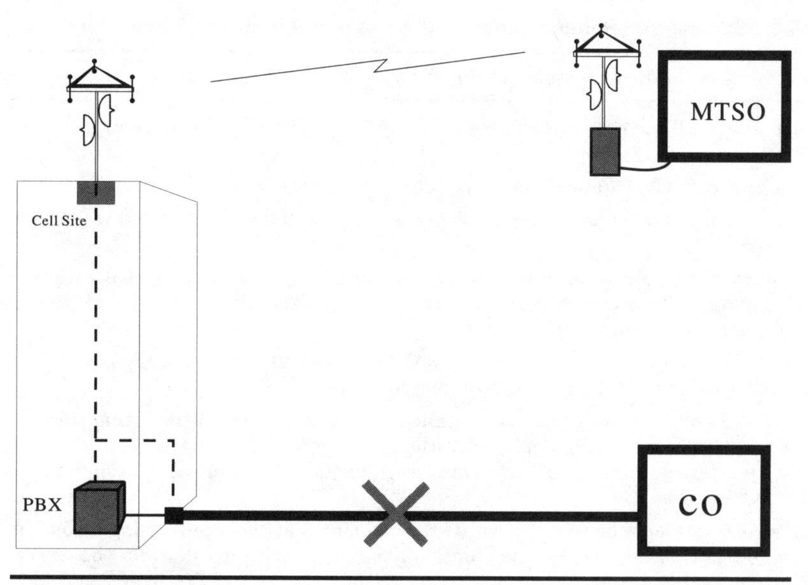

Figure 11.11 Captive cell sites can be installed at the customer location and connected to a PBX. Line sensors will monitor the wires from the CO.

This will likely change when the migration to digital cellular takes place and CDMA or TDMA systems become far more economical. The cell site will have a connection from the top of the building down to the PBX.

Line-sensing equipment will be used to sense that continuity exists to the telephone company's central office. Therefore, the cell site will sit idle in wait mode. Upon sensing that a disruption to the lines (cable cut or whatever) the cell site will immediately come alive and notify the MTSO that all calls should be routed into the cell. This will be done through the intelligent network schemes that are being developed on the SS7 networks. Once the MTSO receives the message, it will process all calls to the cell site located at the customer's premises. Any call being directed toward the wired office will be rerouted across the wireless portion of the network and delivered to the PBX from the cell. The PBX will see the incoming wireless trunk just as any other direct inward dialed call on a DID trunk. The PBX will then process the call to the appropriate extension number whether it is a wired or wireless extension. Only when the cable problems have been rectified will the cell site go back into idle mode. Before going back to idle, the cell site will notify the MTSO that all calls will now be forwarded to the central office for processing across the wired world. The PBX user will never know the difference that this all took

TABLE 11.6 Summary of Requirements for the Use of a Portable Cell Site

Requirements of the cell site at a business
1. The cell site must have the prearranged channel capacity with the carriers in advance.
2. The channel capacity (up to 90) must be fully equipped in the site.
3. Line sensors must be installed on the wires so that the cell can be alerted when to come alive.
4. The PBX must be able to see ringing voltages and dial tone, as well as deal with DTMF signaling. To accomplish this the cards that provide the cellular to telephony interface made by Telular are required.
5. Battery backup on the cell site and the PBX are a must. If not a battery system, then UPS with a mechanical generator should be installed.
6. The SS7 links and network must be installed and tested prior to attempting this connection. SS7 is still being deployed in the local telephone company networks, and it must be extended to the customer's premises for the cell to message through to the MTSO and the CO.
7. Airtime costs will only be incurred while the cell site is actively processing incoming and outgoing calls. There may be a monthly fee by the carrier to allow for the use of the channels while in idle state.
8. The system should be periodically tested to ensure that everything is working according to the design. Too many instances have occurred in the past where a recovery system was never tested and failed to operate as designed when needed the most.

place. There are some caveats in this whole scenario; these are summarized in the Table 11.6 below. These issues must be pre-addressed for the system to function as described above.

The above may seem like overkill for some organizations. Either the costs associated with this temporary stopgap are too high, or the organization is too small to use such a service. This may well fit into an agreement with the owner of a multistory building providing the site for a fee on a shared basis to the tenants of the building. So too these needs may go away as the PCS rolls out to ubiquity, since every phone has the option of being a wireless connection arrangement. Why then spend the money for this arrangement when the network will deliver this form of service as a norm? That is a tougher question that users will have to answer. However, it would be beneficial to remember that the ubiquity of PCS is something that is not slated for a few years. Only spotty capabilities will be available in the onset of the deployment. It is likely that the year 2000 is the year of PCS ubiquity not 1994. How long can the organization wait, is there a stop gap until this happens? The answer is something each organization must deal with individually or collectively.

Using Infrared for Disaster Recovery

For all that has already been stated for the use of radio-based systems, the use of infrared must also be considered when planning for the necessary recovery steps. Infrared does not bear the licensing brunt of the radio-based systems such as the microwave, satellite, or cellular systems. The limitation in infrared is to get clear line of sight and that the use is distance-sensitive. An infrared system during a disaster could be used up to 2 km distances to provide connectivity to a central office in the event of a cable cut. Getting to a clear line of sight to the CO may be the biggest challenge facing the user.

The atmospheric conditions and limitations also must be considered, since this is not a technology that will meet all needs for all organizations. If the site is in an area where line of sight will be a problem, this technology will not necessarily meet the need. So too, if the area is prone to dense fog or heavy shimmer as described in Chap. 3, then another solution may be sought. The use of infrared will, however, work for some where in the event of a cable cut or other disaster the system is portable enough to cart it along and have it set up within an hour. Using the parameters that were discussed earlier in this book, the infrared can deliver up to 96 channels of 64 Kbps voice or data capacities, or a full native LAN speed to link buildings or networks together. From the voice world the interfaces to the infrared system will allow for a standard 4 wire digital transmission service.

In the event a building is impaired beyond occupancy, then the infrared system can be set up quickly to establish a group of circuits from a central office, from another building, or from a POP. If co-location space is not available then a "rusty switch" or "broom closet" approach may suffice. This is a concept where a room or closet space can be rented and the infrared system set up there. An interconnection arrangement to a wired connection can then be established and brought virtually anywhere. This is shown in Fig. 11.12 where a broom closet is used to connect up to four T1 circuits to a remote site. The fact is that this unlicensed technology is particularly attractive to meet this need, since it can be installed quickly and removed or reinstalled at a different location after the disaster is over.

Wireless PBXs in Disaster Recovery Modes

Up to now, the gist of the recovery techniques have dealt with the actual outside circuitry to connect to the outside world. However, if a disaster is isolated to a single building or a single floor within a building, then the possibilities abound. One of the techniques that should be considered to get service back and running is the use of the wire-

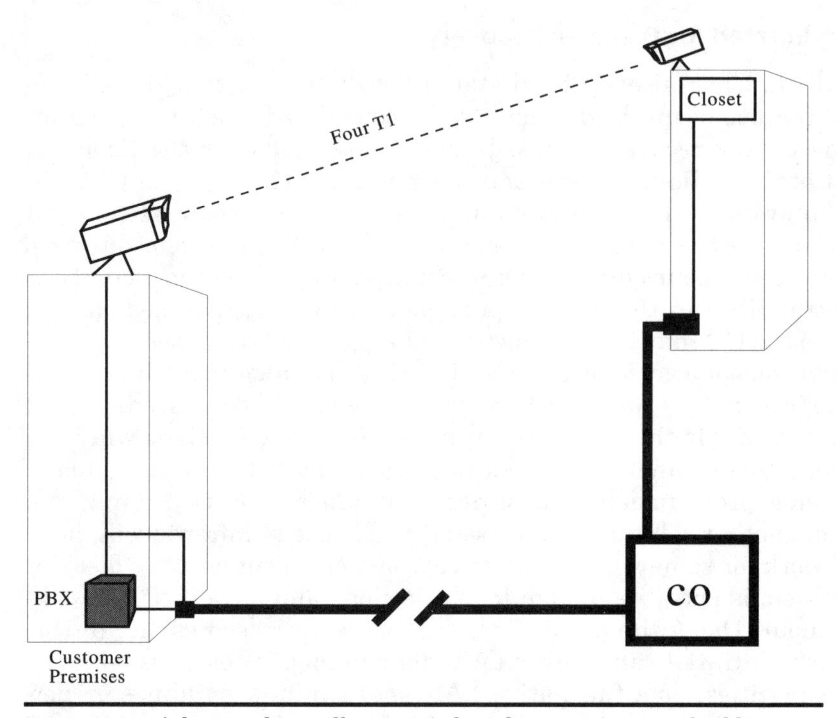

Figure 11.12 A broom closet allows an infrared connection to a building adjacent to the CO if space is not available in the CO.

less PBX. As discussed in Chap. 10, the PBX is not totally wireless, there are always some wires that may be needed to connect to some service or other. In the event a fire, flood, or cable cut impairs the ability to get to a floor or building, a few different scenarios can be played out. These are thought processes that must be preplanned or coordinated as much as possible in the future. However, the PBX with the capability of using a picocell will allow for the quick connection for the use of day-to-day operations.

In this first case, a building may be severely impaired from a fire or flood. Perhaps the building was untouched, but a fire occurred next door. Either way, the organization may not be able to get into the building for some period of time while it is under the control of others, such as the fire department or police department. Therefore, the organization must find a means of serving the user's needs without going through extensive wiring or rewiring in a building. In this case a wireless PBX might be brought up to the building in a van, or it can be moved into the building for that matter. The wireless PBX will require either a wired or wireless connection to the telephone compa-

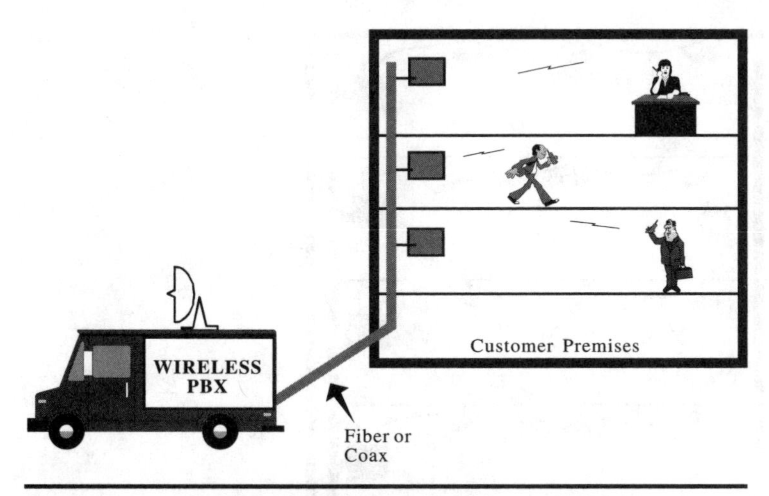

Figure 11.13 A wireless PBX can be rolled up to a building; only limited cabling to picocells would be required rather than prewiring a building. Wireless phones can be used in the interim.

ny office for the dial tone necessary to conduct business. However, wiring stations throughout the building will be both expensive and time-consuming.

The costs of wiring to the stations will be exorbitant and will be "sunk costs," meaning that they are unrecoverable. This is walk-away technology. Instead, the use of picocells on each of the floors will require a small amount of wiring to each of the floors, four wires (two pair) to each picocell will require orders of magnitude less wires than wiring several hundred telephone sets throughout the building. Rather, the picocell can act as the interface for the PBX dial tone to the portable sets and still deliver all of the same features as what the users had in the original system. (See Fig. 11.13.) Since this is possible to deliver in a van, it can be a short-term rental, and it can be rolled away after the need has been met. This will reduce the total overall cost of recovery. PBXs serve more than just dial tone; they also serve many of the facsimile and data needs of the organization. Therefore, a mix and match of the needs can be accommodated through this versatile vehicle. Nothing states in any rule that only one single application can be met.

In the second option, if a major cable is cut inside a building, which just happens to occur every business day, the wireless PBX features can again serve to get service restored to an area quicker than having to splice or rerun cables to every single desk top. In Fig. 11.14 this is accomplished through a few wireless cells being distributed in a

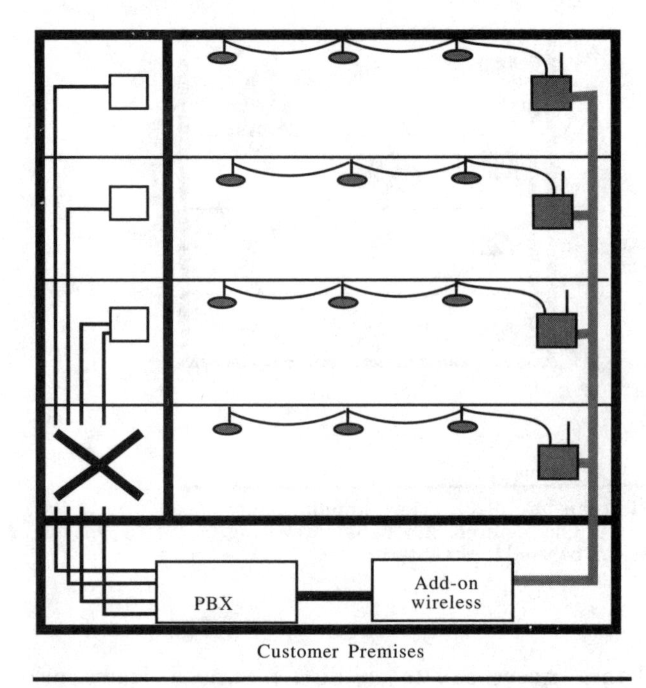

Figure 11.14 In the event of a major cable cut inside a building, a few picocells can be used to support the telephone needs for short terms.

building. Clearly, if there is any EMI or RFI*, then the use of an RF-based system must be thought out. The use of spread spectrum, direct coding, is an alternative to these problems, since most of the other interference will not affect this system extensively. In a medical environment, clean-rooms and heavy factory areas where the environment is full of RFI and EMI, the use of a spread spectrum system is a must.

A fixed frequency radio-based system will likely experience too much disruption, or the radio-based system (in fixed frequency) could cause problems for machines and test/lab equipment in use already. One must therefore be aware of the location that such a system will be used. The use of any such system is always suspect from an interference standpoint. This has already been the case with some medical facilities where the X-Ray, CT Scan labs, and the like have experi-

*EMI stands for electromagnetic interference, and RFI stands for the radio frequency interference. Both can cause significant problems with the transmission of information on a radio frequency system.

enced difficulty coexisting with radio-based systems. Further, in many campus type environments, the users have experienced loss of calls or static in areas where the signal is being disrupted. To overcome this particular problem, picocells must be positioned closer to the end user. However, when the interference problem reigns, just putting more cells around only adds to the problem rather than solving it. Care must therefore be taken in the selection and design of a wireless system. The end user will need assurances from the provider that the system will not cause this problem in the environment and that if it does cause the problem, it will be removed at no charge to the end user. This is merely a cautionary statement, rather than a condemnation of the entire technology.

Wireless E-Mail (Messaging) in Disaster Recovery

For every need there appears to be a solution that can be applied. The use of various techniques in a disaster recovery mode is to be assured that something in the closest form of normalcy can be provided to supplement services that have been impaired for whatever reason. Many organizations have always viewed electronic messaging systems as administrative tools to assist the user population get their jobs done more efficiently. This was true when the use of E-mail was first introduced. It was seen as a neat service to send messages back and forth between and among the organization, especially when the printing option is available. A written record can be created of the mail between users, something that could not be done with voice messages. Therefore, the use of E-mail as an administrative tool was accepted. Yet, over time more organizations have learned that the routine day-to-day chores that are fundamental to running the organization have been systematically added or tied to the E-mail system. What was once just a casual service that people could use rather than speak to each other has become a critical asset in the organization. The use of the E-mail is a mission supporting device now. Thus, when planning a disaster recovery mode transition, the use of wireless messaging has found its way into the niche. In Table 11.7 the reasons that the E-mail service has become so popular are highlighted. This is a representation of why the service is viewed as critical, but also may provide a better understanding of how it may be used in the future should a disaster strike.

Clearly then the use of an electronic messaging system has become readily acceptable within the organization. To provide for a quick connection then, the users could be provided with a tool that they have already become adjusted to and feel strongly about. In most organiza-

TABLE 11.7 A Summary of Why the Importance of E-Mail Has Developed

E-mail as a critical resource
1. Users do not want to talk to other users, they perceive the verbal interaction as a waste of precious time.
2. Many organizations use the E-mail service for quick broadcast messages to all users, or groups of users, which requires only that the message be sent once to assure that all users receive it.
3. E-mail allows for a "return receipt" or audit trail that is more difficult to implement and administer than in the voice world. Particular attention is paid to the date and time stamps that can be obtained for audit or record keeping purposes.
4. Many users feel that they must have the ability to print down the messages that they receive. This is something that they cannot do with the voice mail or real-time voice messages. This print service allows the users to keep accountability of certain projects, etc.
5. Many organizations have become adjusted to working with the dynamics of time zone differences between users in the same organization, or with major geographic distances separating members of the same work group. Therefore, the use of a written message allows some better flexibility in dealing with individuals in different languages (countries) or in various time zones.
6. Some have realized that file transfer capabilities that are not yet provided (between CAD systems, as an example) can become a bottleneck. To overcome the limitations of this lack of transfer service, the end users have discovered that a binary file can be appended to a mail message and sent through the organization. This may not be the best use of a system, but it certainly works. Therefore, the use of E-mail as a file transfer vehicle becomes critical.

tions the use of E-mail is tied to a host-based system or to a LAN environment, with hooks to the outside world through some dial-up arrangement. This could be through such services as CompuServe, Prodigy, or AT&T Mail. These are based on a wired perspective. Regardless of the mobility issue the user must typically find a telephone device and plug a modem into the wired system to communicate the mail between or among each other. This works; there is no doubt about the reliability of this service. However, if a situation develops that users will be very mobile, or displaced from their primary location, then the dependency of finding a wired communication gets in the way. To overcome this portion of the limitation the use of wireless messaging such as the ARDIS or the Mobitex services discussed in Chap. 10 will do.

One such situation where wireless messaging was used effectively was how the Red Cross dealt with the aftermath of Hurricane Andrew in Florida in the late summer of 1992. IBM personnel, associated with the mobile data systems group, contributed 30 IBM PC radios and the rights to use the ARDIS network to the American Red Cross in support of the relief efforts. Using the current speeds of up to

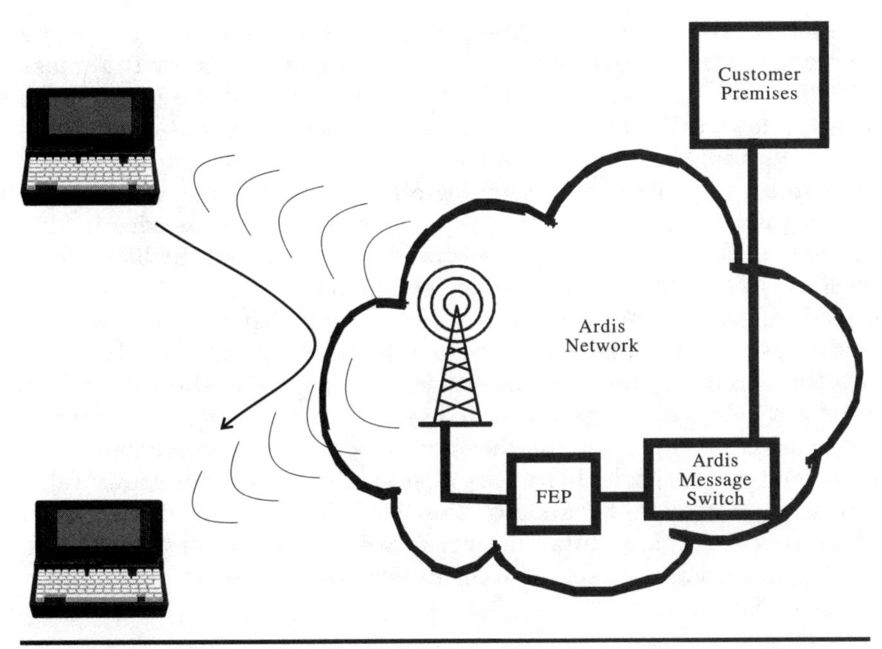

Figure 11.15 The wireless messaging system allowed customers access from remote PCs to the host for easy access to files, information on food, shelter, etc.

19.2 Kbps on the network for routine short administrative messages and access to host systems for file transfers, the Red Cross was able to quickly learn how to use the system and immediately began the dissemination of information to as many as 15 service centers in the disaster struck areas. The ARDIS system was initially designed to relay information to the relief crews about the status of food, medical supplies, and shelter facilities that were being provided. In Fig. 11.15 a representation of how the connections worked is shown. This allows for instantaneous communications and information swapping without bogging down the process with long hold times, eliminated confusion, and overall provided the Red Cross with the information that was critical to supporting and saving lives. After any disaster the demand for a communications service will be exceptional.

A calculated guess is that the need for communications after a disaster is tenfold more than before such an event happens. The reason is obvious that the interchange of information for status reporting is increased exponentially because of the coordination efforts required. The Red Cross was no different; the demand for communications capabilities were dramatic during the aftermath of the devastation caused by the hurricane. The system allowed the Red Cross personnel to coordinate location of supplies and shelter, dynamically move these supplies as necessary, or redirect victims who needed shelter to the appropriate location without confusion. This is not to imply that mass confusion was not evident, but for the E-mail side of the picture, the system allowed the workers easy access without long dial-up and log-on procedures. The Red Cross logged on in the morning only. This prevented the protracted log-on/log-off procedures that would have been required with dial-up or leased lines. The phone service in this area was also impaired because of the effects of the storm. Therefore, the Red Cross felt that the use of telephone service would be hit or miss. Lines were downed and needed repair. Other locations were flooded, causing the disruption of the telephone service. This happens with the wired communications services after a natural disaster. The use of a wireless connection is a little easier to transition from during the disaster to the "after the disaster" mode. In this particular case, the use of wireless telephony was also available through the cellular networks. However, because of the total destruction of the wired infrastructure, the cellular networks were heavily congested. The messaging through packets of information were easier to deal with in a real-time need.

Wireless LANs in a Disaster Recovery Mode

As more organizations proliferate the use of local area networks in their day-to-day use of computing resources, a new problem has surfaced. Problems exist as follows:

- The technology is somewhat complex to manage from the perspective of a LAN manager/administrator's daily duties. Users continually demand more services and applications on the networks, and they develop newer dependencies on the ability to access and manipulate their data on the fly.
- The users are continually adding applications that could become a problem on a network, either through a configuration of their system or through a conflict in the applications that they install on an already existing network.
- The use of wires throughout the building or enterprise complicates the administration and management of the overall LAN. Wires break, humans pull the wires out of the wall, and other situations—such as power spikes and dips—cause the hardware to fail.

These are not disastrous situations, merely management problems that will be in the way of providing high-quality service to the user population. However, adding to this is the fact that the cumulative effect of downtime has sent a pessimistic note out to the industry.

LANs fail on an average of 2–3 times per month and the average amount of downtime is approaching $50,000 per hour. Cumulatively this indicates that the amount of time invested in the maintenance of a LAN is reaching disastrous proportions. An interesting statistic from the studies conducted on the causes of downtime indicates that 65–70 percent of all LAN downtime is attributed to *cabling* problems.

The use of wires in the office environment allows for the high-speed throughput but also accounts for the disruptions that go along with the running of a LAN. How does one fix that problem? Clearly, the solution is to be aware of the implications of poor installation practices. Newer methods and tools are available to provide for a wiring solution to resolve some of the causes of this amount of downtime. But, that is only part of the problem. To overcome the possible ramifications of downtime as a result of wiring, the use of wireless LAN services is now rising to the surface. In general, the wireless LAN is not as fast as a wired system, yet. As the wireless standards that have recently been agreed to in the industry to speed up the throughput and as the access contention methods are improved to prevent collisions, then the wireless LANs may find a new niche.

However, in the event of a catastrophic loss of the building, the possibility of fires and floods or bombings in buildings, and the likelihood of major downtime situations due to the wiring systems, wireless LANs can be used as a short term solution. The use of an alternative building for example, can provide an organization with a quick fix to a failure in the building. Some of the scenarios that may have to be dealt with are summarized in Table 11.8 as a simple check-off of what to plan for.

In the event any of the above happens, the organization may be able to recover quickly, or the recovery process may be extended over longer periods of time depending on the overall situation. An example of this would be to consider flooding. In most cases the attitude is that a simple fix can be put in place, which very well may be true. However, if one were to consider the problems that were experienced

TABLE 11.8 Scenarios That Can Lead to a LAN Disaster

Possible scenarios that will cause LAN disasters
Total loss of building
Flooding in a closet
Backbone cable cut
Loss of power to building for extended periods
Fire in a closet or in a server room
Loss of cable records

TABLE 11.9 The Possible Steps of Repairing or Replacing Cables

Sequence of events that should be looked at after a disaster
1. Assess the damage; determine if the site is reusable
2. Determine if the cabling can be reused or salvaged
3. If site is unusable, find new location
4. If site is usable, remove old wires, blocks, and debris
5. Get contractors to pull new backbone wires (in either site)
6. Install new horizontal station cables
7. Reconnect the systems to the cabling
8. Test cable for continuity, shorts, grounds, etc.

in the summer of 1993 in the mid-western portion of the United States, the flooding that ravaged this portion of the country was for weeks and the results could take years to overcome. Consider the ramifications of a flood that has totally destroyed the building and the associated equipment. The next steps to take are summarized in Table 11.9. This is not an all-inclusive list of events, but a physical listing of the areas that would be considered.

The reasoning for Table 11.9 is that once an estimate of the damage has been made, the next step involves the demolition of the old wiring system or the physical relocation to a new building. In either case, the wiring of a building for LAN connectivity will take a considerable amount of time and expense. What about the possibility of having to move into temporary quarters for a period of 30 days? To wire the LAN users' terminals for a short period of 30 days still will require the same expense for the wires; then the wired costs are sunk. Once the time comes to remodel the old space, the rewiring will have to be done. Now the wiring has been bought twice, and the service is no different for all this expense.

The sunk costs are only a part of the problem. There is still the timing issue while users are out of service while the cabling is being installed and tested. Overall this can be an expensive solution for a short-term fix. In a wireless world the cabling can be supplemented with a radio-based system or a light-based system such as those discussed in Chap. 9. The use of a wireless LAN can be used in the event of a backbone cable cut, as shown in Fig. 11.16, to get the department up and running quickly. Or as an alternative the wireless LAN can be used in the event a new building is required. This will prevent the need for a total wiring of the building when a single cell or departmental wireless LAN can be used. In Fig. 11.17 this is shown as a possible solution. The costs for such a service may appear to be extraordinary at first glance. However, one should consider that a move to

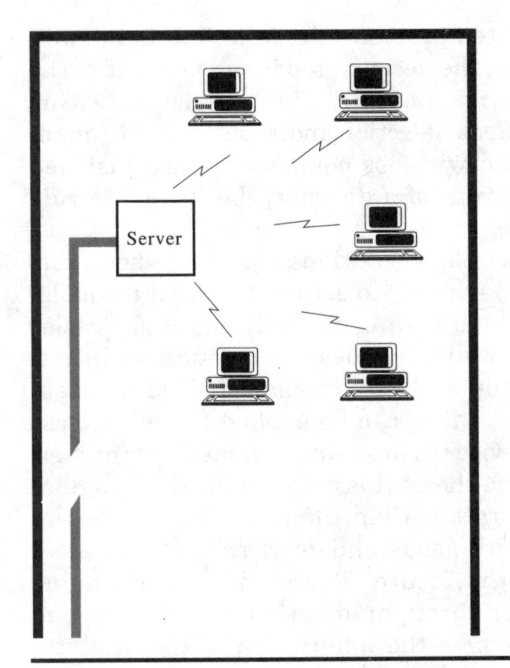

Figure 11.16 Wireless LANs can be used after a backbone cable cut in the same building.

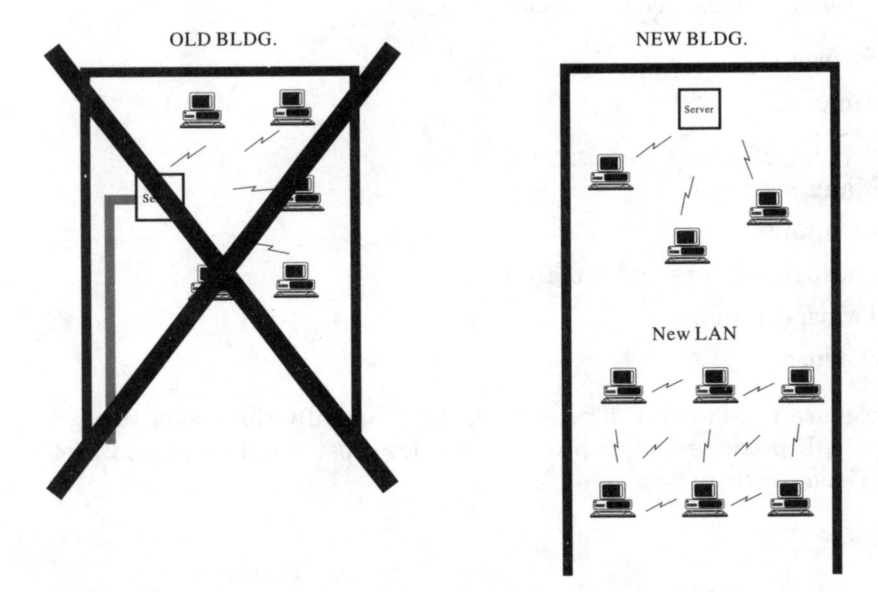

Figure 11.17 Wireless LANs used in a new building prevent the sunk cost of wiring several LANs. These can be installed much quicker than new cables.

a new building requires a two-step process—the first step is to move into and wire the temporary site, the second step is to move out of the temporary site back into the permanent site. This is a double wiring cost that would make the wireless selection more palatable to management. The actual cost of the wireless components are justified after one to two moves. In the event of a disaster, the two-move rule is met easily, as described above.

All in all, every one of the wireless solutions that have been discussed are capable of being quickly put into action. Licensed technologies take a little longer to plan and install; the unlicensed are easier to install on the fly. However, both techniques can be effective in getting an organization's business up and running quickly and inexpensively. Many other situations exist that can be applied to the wireless world, the quick and usable services. These are obviously too numerous to list in this book. However, the real issue is to let the imagination be the only limiting factor rather than the technology. The vendors are coming along nicely with goods and services that will meet the communications needs of the future. The standards are being hammered out to deliver higher throughput and more security and reliability of the products. It is now the application of the products and standards that will have to catch up. Some of the examples used to make a point in this book have been geared to opening the mind to opportunities. Following are industries that will clearly benefit from the use of wireless technologies:

- Medical
- Financial
- Retail
- Manufacturing
- Computer
- Package delivery and handling
- Transportation
- Defense

These are not listed in a priority order. Hopefully the use of the systems will produce higher productivity, less costs, and increased profits. Good luck!

Index

ABOUT THE AUTHOR

Regis J. (Bud) Bates, Jr., has more than 27 years of experience in the telecommunications and management information systems, with specific experience in end-user management, systems integration, disaster avoidance and recovery planning, strategic planning, and cost containment or reduction. Mr. Bates is the president of TC International Consulting, Inc., in Phoenix, Arizona—a management consulting firm specializing in voice and data, LAN communications services and implementation, technical training development and instruction, and consulting services at high levels of domestic and international organizations. He is an active lecturer (having presented to nearly 15,000 users) and author of numerous articles and books, including *Disaster Recovery Planning: Networks, Telecommunications and Data Communications* (McGraw-Hill, 1990), *Disaster Recovery for LANs: A Planning and Action Guide* (McGraw-Hill, 1993); and *Introduction to T1/T3 Networking* (Artech House, 1992).

ABOUT THE SERIES

McGraw-Hill Series on Computer Communications is McGraw-Hill's primary vehicle for providing communications professionals with timely concepts, solutions, and applications. Jay Ranade, series advisor and editor in chief of J. Ranade IBM, DEC, and Workstation Series has more than 125 published titles in various series.